Pocket Guide Geology in the Field

Tom McCann

Pocket Guide Geology in the Field

Tom McCann
Steinmann-Institut für Geologie
Universität Bonn
Bonn, Nordrhein-Westfalen
Germany

ISBN 978-3-662-63081-5 ISBN 978-3-662-63082-2 (eBook)
https://doi.org/10.1007/978-3-662-63082-2

This book is a translation of the original German edition „Pocket Guide Geologie im Gelände" by McCann, Tom, published by Springer-Verlag GmbH, DE in 2019. The translation was done with the help of artificial intelligence (machine translation by the service DeepL.com). A subsequent human revision was done primarily in terms of content, so that the book will read stylistically differently from a conventional translation. Springer Nature works continuously to further the development of tools for the production of books and on the related technologies to support the authors.

Responsible Editor: Simon Shah-Rohlfs
This Springer imprint is published by the registered company Springer-Verlag GmbH, DE part of Springer Nature.
The registered company address is: Heidelberger Platz 3, 14197 Berlin, Germany

Contents

Overview

T. McCann, *Pocket Guide Geology in the Field*,
https://doi.org/10.1007/978-3-662-63082-2_1

Minerals are naturally occurring inorganic solids with a defined chemical composition and a specific crystal structure. They are the basic components of the Earth as well as other celestial bodies (e.g., the Moon, meteorites).

Rocks are natural and stable aggregates or combinations of one or more minerals that can be divided into three main groups—magmatic, sedimentary, and metamorphic (◘ Fig. 1.1, ◘ Table 1.1).

Magmatic rocks are formed by the cooling of molten or partially molten material (magma) on or within the Earth's crust. Extrusive magmatic rocks (e.g., basalts) form when magma cools on, or close to the surface, while intrusive magmatic rocks (e.g., granites) form as a result of cooling within the Earth. During the cooling process, characteristic minerals, and mineral series, are formed (◘ Fig. 1.2).

Sedimentary rocks are formed either by the consolidation and cementation of loose sediments (e.g., sands) or organic matter (e.g., coal) that were deposited in layers on the Earth's surface, or as a result of chemical precipitation (e.g., carbonates, evaporites) (◘ Fig. 1.2).

Metamorphic rocks are formed from existing rocks that are transformed as a result of changing temperatures and pressures. These new conditions result in mineralogical, chemical, and structural changes.

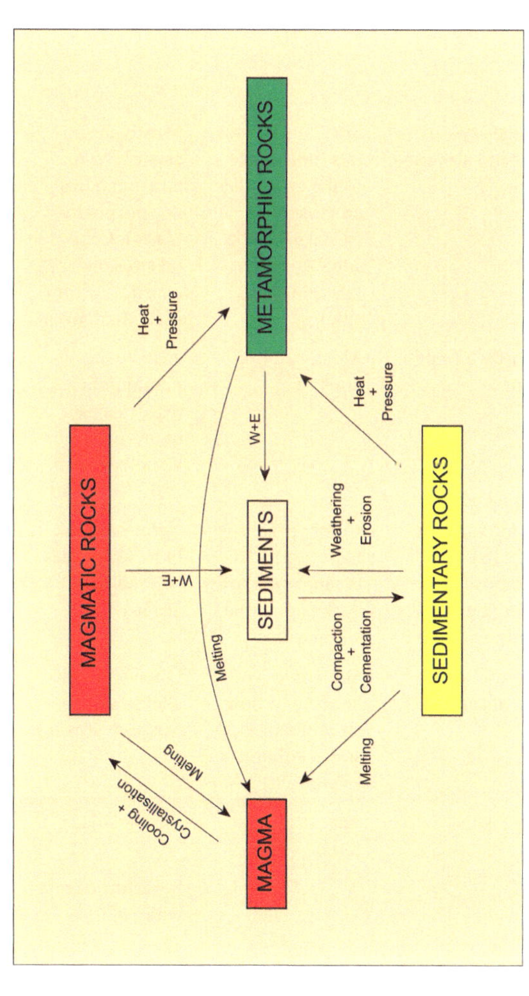

Fig. 1.1 The rock cycle (W: weathering; E: erosion). (After McCann and Valdivia Manchego 2015)

1

Table 1.1 General properties of magmatic, metamorphic, and sedimentary rocks. (After McCann and Valdivia Manchego 2015)

	Plutonic rocks	Volcanic rocks	Metamorphic rocks	Sedimentary rocks
Crystallinity	Crystalline	Crystalline	Crystalline	Mainly comprise fragments in a matrix/cement; some limestones and evaporites are crystalline
Size of crystals/fragments	Large crystals; mineral size variable	Small crystals (not visible to the eye; microcrystalline to glassy), with some large crystals (porphyritic)	Mostly large crystals (some larger crystals, i.e., porphyroblasts). Crystal sizes may be constant within individual layers	Fragments (clasts/grains) can be very variable (e.g., sandstone, conglomerate)
Composition	Mostly ≥2 minerals	Mostly ≥2 minerals	May be monomineralic (e.g., marble, quartzite), but usually ≥2 minerals	May be monomineralic (e.g., limestone, dolomite), but usually ≥2 minerals
Color	Color variable—light (e.g., acidic composition) or dark (e.g., basic composition)	Color variable—light (e.g., acidic composition) or dark (e.g., basic composition)	Color variable—sometimes banded (e.g., gneiss)	Color very variable
Structures	Normally no stratification	Sometimes with layering or flow structures. Occasional columnar jointing	Often with parallel structure (e.g., schistosity)	Mostly pronounced stratification/bedding
Fossils present	No fossils	Rare fossils (e.g., tuffs)	Rare fossils	Often contain fossils
Reaction with HCl			Sometimes react with HCl	Limestones react strongly with HCl; weaker reaction in dolomites

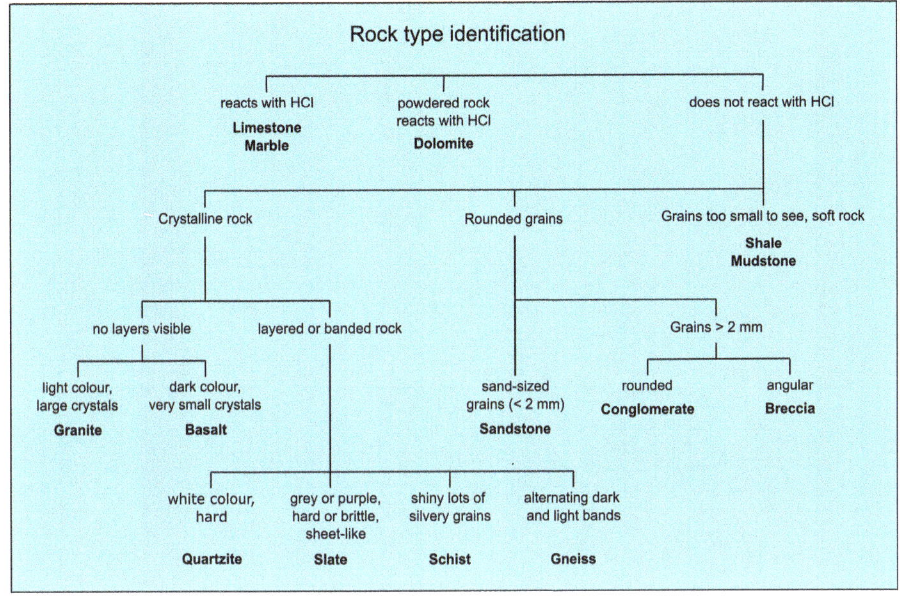

Fig. 1.2 Rock identification diagram

Minerals

Contents

T. McCann, *Pocket Guide Geology in the Field*,
https://doi.org/10.1007/978-3-662-63082-2_2

2

Table 2.1 Frequency (by vol.) of minerals in the Earth's crust (After Ronov and Yaroshewsky 1969)

Mineral	Percent by volume
Plagioclase	39
Alkali feldspars	12
Quartz	12
Pyroxene	11
Amphibole	5
Mica	5
Olivine	3
Clay minerals (+chlorite)	4.5
Calcite (+aragonite)	1.5
Magnetite (and titanomagnetite)	1.5
Dolomite	0.5
Others (garnet, kyanite, sillimanite, apatite, etc.)	4.9

While there are c. 4600 recognized minerals, only around 40 of these, including quartz, feldspar, mica, pyroxene, amphibole, and olivine, are common rock-forming minerals. Other important minerals include calcite, dolomite, magnetite, pyrite, chlorite, clay minerals, epidote, magnetite, and hematite (Table 2.1). The rock-forming minerals can be divided into 3 groups:

- **Primary/Key minerals** (>10% by volume): quartz, feldspar, pyroxene, amphibole, biotite—these are the main components of the rock
- **Accessory minerals** (1–10% by volume): zircon, apatite, titanite, tourmaline etc.—although present in small amounts they can provide valuable information with regard to the genesis of the rock

2.1 Crystals

Every crystal has a defined structure (crystal lattice) resulting from the spatial arrangement of atoms or ions within a mineral. The lattice structure, which determines the geometric form of a crystal, also ensures that their chemical and physical properties are relatively uniform (except in zoned minerals). Differences in crystal structure and/or chemistry can greatly influence mineral properties (e.g., graphite vs diamond).

Mineral description and identification is based on a range of factors, including their crystallographic, physical and chemical properties, as well as their field occurrence.

2.1.1 Crystal Symmetry and Crystal Systems

On the basis of their symmetry, crystals can be grouped into seven different systems (◘ Fig. 2.1). Minerals within each crystal system show particular habits (i.e., general shape).

- **Cubic system**—basic cube form, including e.g., octahedron, rhombic dodecahedron (e.g., galena, pyrite)
- **Tetragonal system**—similar to the cubic system but with a longer axis forming e.g., prisms and pyramids (e.g., chalcopyrite, rutile, zircon)
- **Orthorhombic system**—similar to the crystals of the tetragonal system, except for the lack of a square cross section (e.g., olivine)
- **Hexagonal system**—6-sided prisms with a hexagonal cross section (e.g., apatite, beryl)
- **Trigonal system**—crystals of this system have a threefold axis of rotation instead of sixfold as in the hexagonal system. In addition, the cross section of the prismatic basic form is triangular unlike the 6-sided cross section in the hexagonal system (e.g., calcite, dolomite, hematite, corundum, quartz).
- **Monoclinic system**—like an oblique tetragonal system; two axes are vertical, the third is oblique (e.g., augite, epidote, gypsum, hornblende, orthoclase)
- **Triclinic system**—usually not symmetrical from one surface to the other, because all three crystal axes have different lengths and none are at right angles to the others (e.g., microcline, plagioclase)

2.1.2 Crystal Habit

The characteristic shape of a crystal (crystal habit) can be very useful for the identification of minerals, for example, garnets are often granular, while micas are platy/foliated and amphiboles can be needle-like. Sometimes two crystals of the same mineral grow together to form **twins** (◘ Fig. 2.2).

When identifying minerals in **hand specimen** it is particularly important to note the presence of crystal faces. These can be divided into three groups:

- **Euhedral/Idiomorphic**—mineral grains with well-developed crystal faces
- **Subhedral/Hypidiomorphic**—mineral grains are partially enclosed by crystal faces
- **Anhedral/Xenomorphic**—mineral grains without any visible crystal surfaces

Particular habits (crystal shapes) can be useful for the precise identification of mineral grains (◘ Fig. 2.3):

- **Granular** or **spherical**—individual crystals are roughly equidimensional or spheroidal, e.g., garnet, fluorite.
- **Tabular, lamellar, platy,** or **foliated**—flat and tablet- or plate-like crystals, e.g., orthoclase (tabular), mica (foliated) and chlorite.
- **Prismatic, columnar, bladed, acicular/needle-like,** or **fibrous**—rod-shaped crystals, including, tourmaline (acicular prisms), gypsum and calcite (columnar),

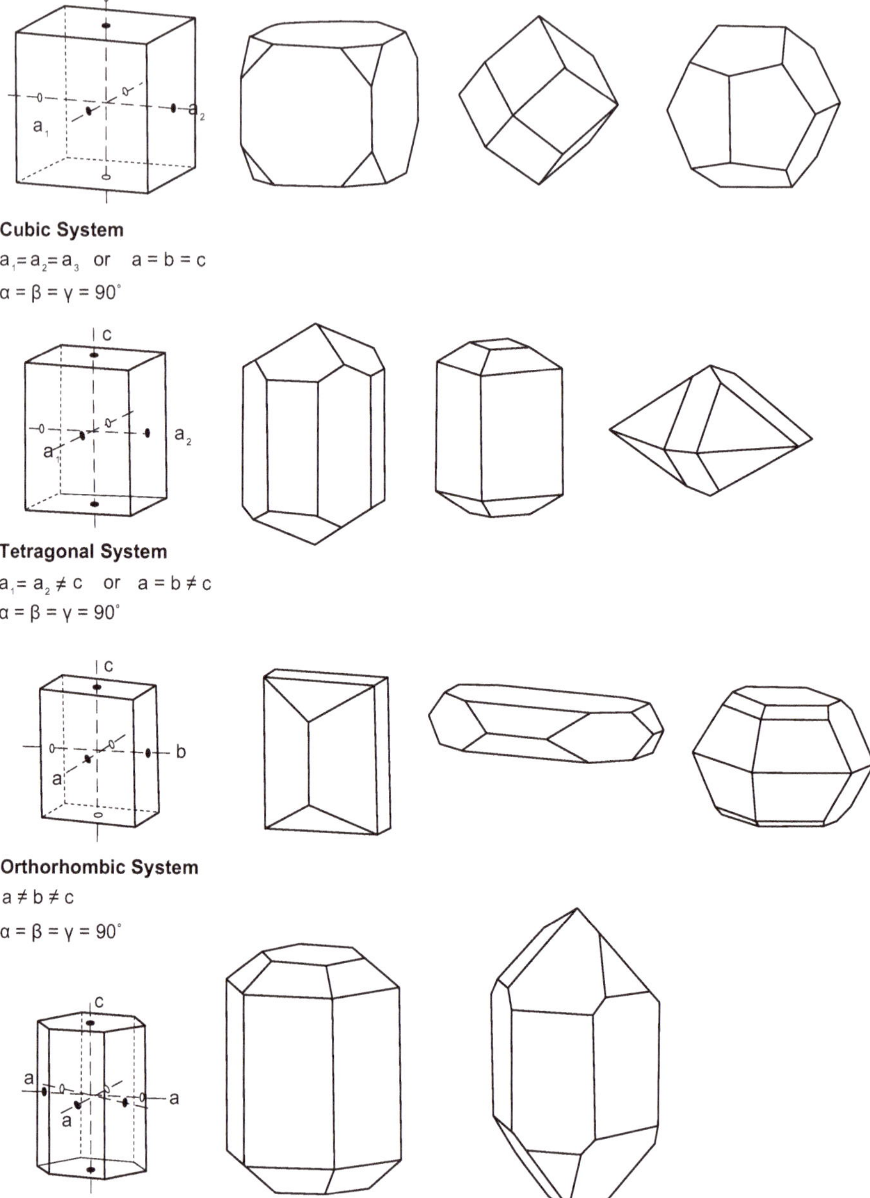

Fig. 2.1 Reference axes and crystallographic parameters of the seven crystal systems and some examples for each system. (After Hamilton et al. 1974; Markl 2004)

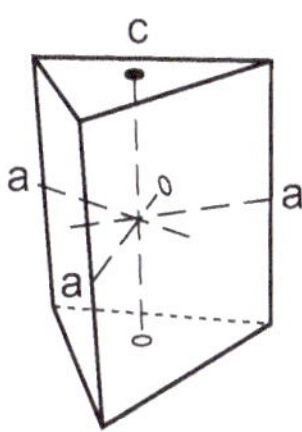

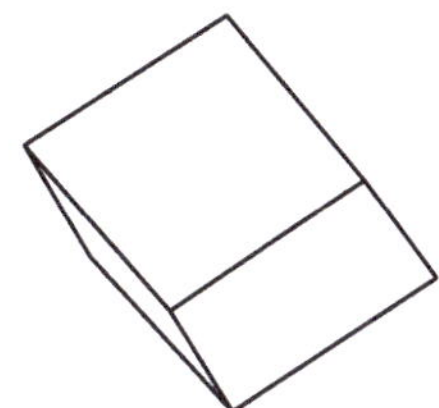

Triagonal System

$a_1 = a_2 = a_3 \neq c$

Angle between a_1 and a_2 and a_3 (γ)= 120°

Angle between a_1, a_2, a_3 und c = 120°

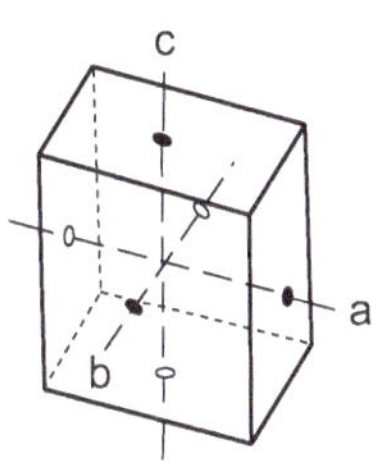

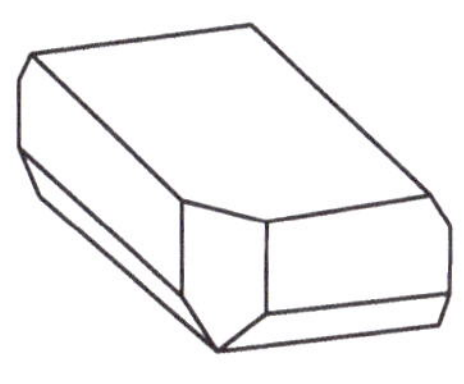
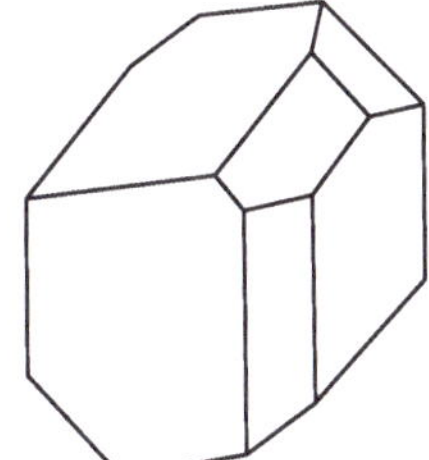

Monoclinic System

$a \neq b \neq c$

$\alpha = \gamma = 90°$, $\beta \neq 90°$

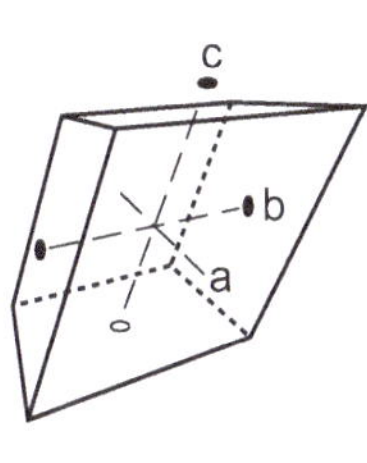

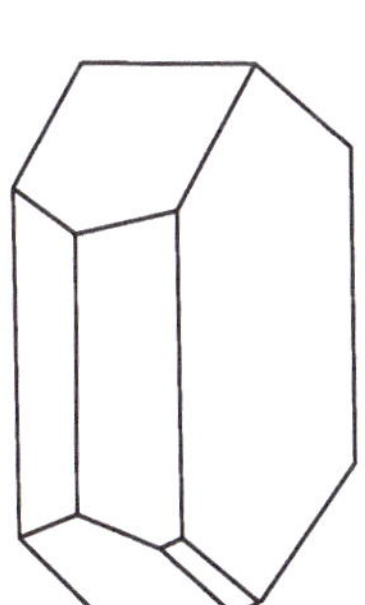
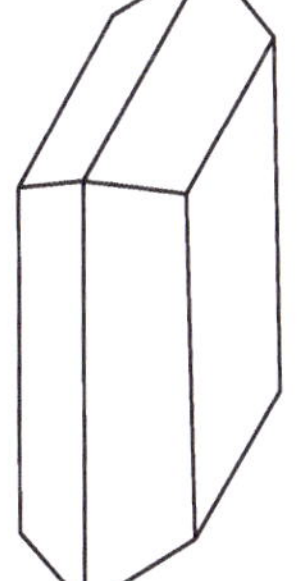
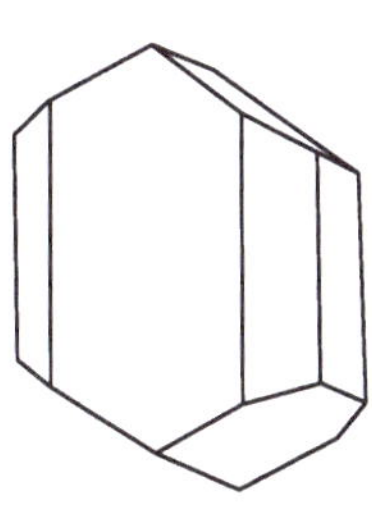

Triclinic System

$a \neq b \neq c$

$\alpha \neq \beta \neq \gamma$

α = Angle between b and c

β = Angle between a and c

γ = Angle between a and b

Fig. 2.1 (continued)

topaz, rutile and amphibole (all prismatic), kyanite (bladed) and chrysotile (fibrous).

- **Dendritic**—branched crystals, e.g., copper, manganese.

2

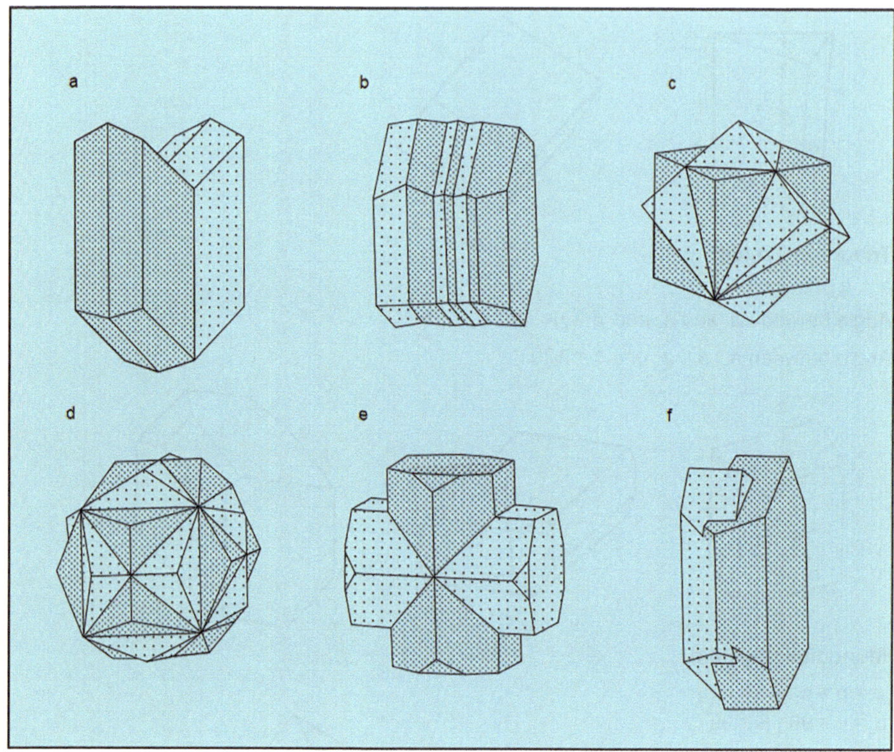

Fig. 2.2 Twin forms in minerals. **a** Swallowtail in gypsum. **b** Lamellar albite twinning in plagioclase. **c** Twin interpenetration in fluorite with cubic morphology, **d** Twin interpenetration (iron cross) in dodecahedral pyrite, **e** Cruciform twinning in staurolite. **f** Carlsbad twining in orthoclase. (After Wenk and Bulakh 2004)

Some minerals also exist as crystal aggregates. While it may not be possible to recognise individual crystals in these aggregates, the overall shape of the aggregate may be diagnostic, for example, **botryoidal** or **mamillary** (e.g., hematite, malachite), or **radiating (e.g., millerite)** (Fig. 2.4).

2.2 Mineral Recognition

Rocks are mineral aggregates, with the individual minerals being in direct contact with one another. In general, rocks—especially those formed by magmatic and metamorphic processes—are more or less in chemical equilibrium at the time of formation. Therefore, the minerals or groups of minerals present within the rock are linked by their shared genesis, as well as by their chemical relationship with one another. Thus, the identity of an unknown mineral can be partially determined by the presence of another mineral (Table 2.2).

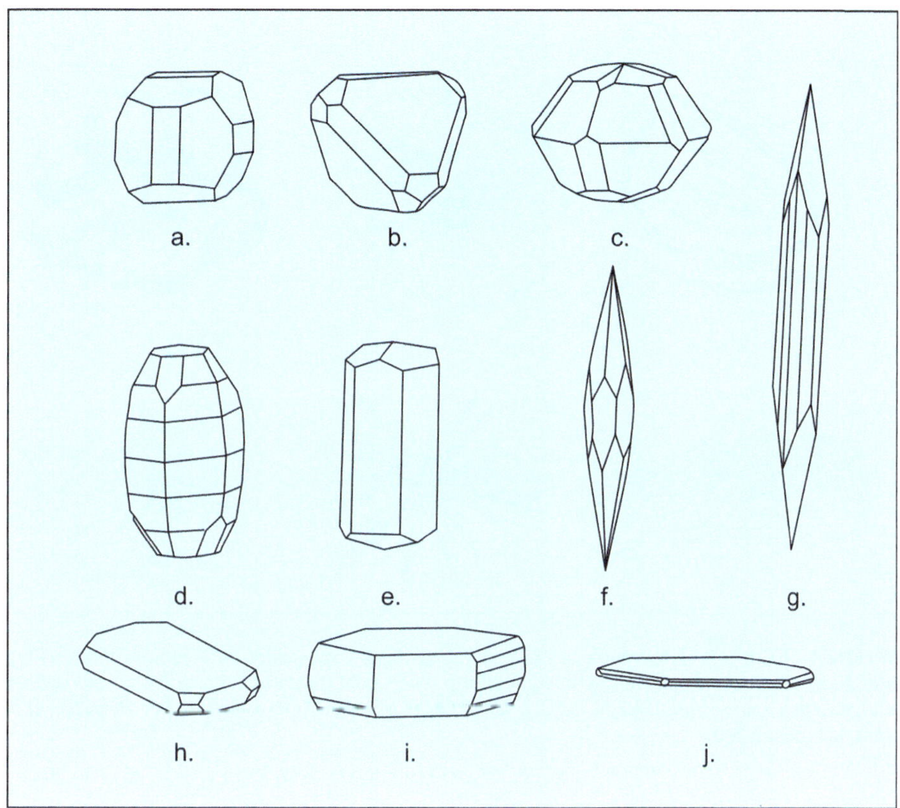

Fig. 2.3 Crystal habits: **a–c** Equant: **a** Pyrite with dodecahedron and cube. **b** Sphalerite with dominant tetrahedron. **c** Equiaxial hematite; **d–g** Prismatic/Elongate: **d** Barrel-shaped corundum. **e** Prismatic calcite. **f** Acicular hematite. **g** Acicular stibnite; **h–j** Flattened and tabular: **h** Tabular orthoclase. **i** Platy muscovite. **j** Platy hematite. (After Wenk and Bulakh 2004)

Macroscopic determination of minerals in the field is done with simple tools such as a hand lens, a hardness scale, a streak plate (unglazed porcelain), a magnet and dilute (10%) hydrochloric acid (HCl). Mineral recognition begins with the determination of the luster—either **metallic** or **non-metallic**. The next stage is to classify the various minerals in each of these main groups (metallic/non-metallic) according to their hardness and color (Figs. 2.5 and 2.6). Finally, other diagnostic properties can be used to confirm the identification (Fig. 2.7).

2.2.1 Optical Properties

The optical properties of a mineral are dependent on the interaction of light with the mineral.

2

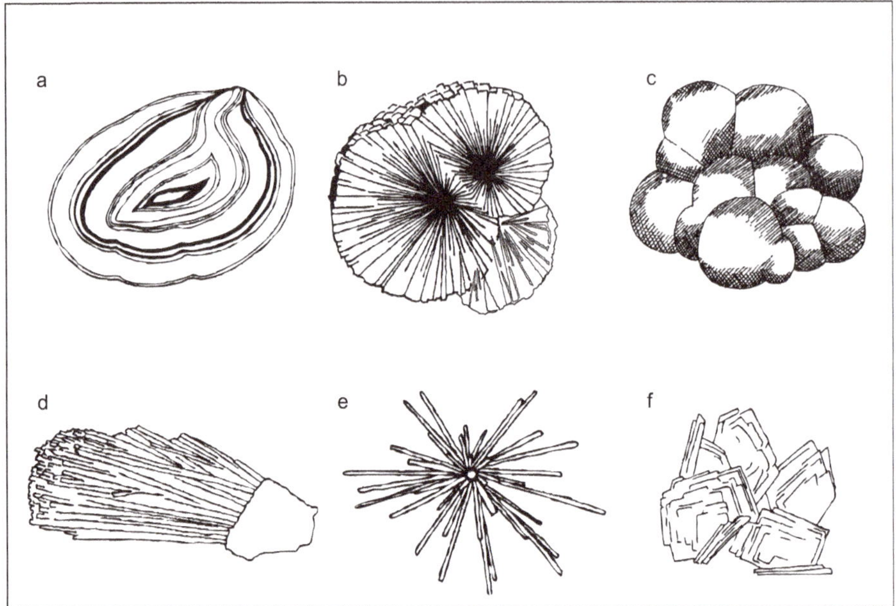

Fig. 2.4 Examples of morphologies in mineral aggregates. **a** rounded with concentric layers (agate). **b** radial mineral aggregate (pyrite). **c** reniform, botryoidal (hematite). **d** parallel, fibrous (actinolite). **e** radial, acicular (aragonite). **f**. platy, scaly, foliated laminae (muscovite). (After Hamilton et al. 1974; Schumann 2007)

Luster

The luster of a mineral is determined by the amount of light reflected from its surface. The main classification, as mentioned above, is metallic (e.g., mercury, copper, gold, pyrite), which have high reflectivity, submetallic (similar to metallic, but duller and less reflective), or non-metallic. The luster of non-metallic minerals is described below:

- **Adamantine luster**—zircon, sphalerite, diamond
- **Waxy/greasy/resinous luster**—apatite, nepheline, halite, plaster, talc
- **Vitreous luster**—quartz, amphibole, pyroxene, olivine, feldspar, baryte, anhydrite, dolomite, calcite, kyanite, epidote
- **Pearly luster**—gypsum
- **Silky luster**—asbestos
- **Earthy/dull luster**—goethite, clay minerals, hematite, chlorite

Transparency/Diaphaneity

Transparency is a measure of the amount of light absorbed by a mineral. Minerals can therefore be classified as **transparent** (e.g., calcite, chlorite, corundum), **translucent** (e.g., mica) or **opaque** (e.g., pyrite).

Table 2.2 Important igneous and metamorphic minerals. (After Blatt et al. 2006)

Minerals in magmatites	Minerals in metamorphic rocks
Silicon dioxide	
Quartz	Quartz
Alkali feldspars	
Sanidine, orthoclase, microcline	Sanidine, orthoclase
Plagioclase feldspars	
Albite, anorthite, plagioclase	Albite, anorthite, plagioclase
Orthosilicates	
Olivine, garnet, titanite, epidote, zircon, topaz	Olivine, garnet, staurolite, chloritoid, titanite, epidote, zircon, topaz, kyanite
Pyroxenes	
Orthopyroxene, clinopyroxene	Orthopyroxene, clinopyroxene
Amphiboles	
Hornblende	Hornblende
Sheet silicates	
Muscovite, biotite	Muscovite, biotite, chlorite, serpentine
Ring silicates	
Tourmaline, cordierite	Tourmaline, cordierite
Oxides	
Spinel, magnetite, hematite, ilmenite	Spinel, hematite, ilmenite
Sulfides	
Pyrite, chalcopyrite	Pyrite, chalcopyrite
Other non-silicate minerals	
	Apatite, calcite, magnesite, dolomite, diamond

Color

Most minerals can have a range of different colors (Table 2.3). For example, garnet can be red, yellow, colorless, and even black. In general, color alone is not characteristic enough to uniquely identify a mineral.

Streak

Streak is the color of the fine powder produced when a mineral is rubbed across an unglazed porcelain plate. The color may be characteristic (Table 2.4). It is particularly useful for opaque ore minerals (e.g., sulfides, oxides), especially those with metallic or submetallic lusters.

2

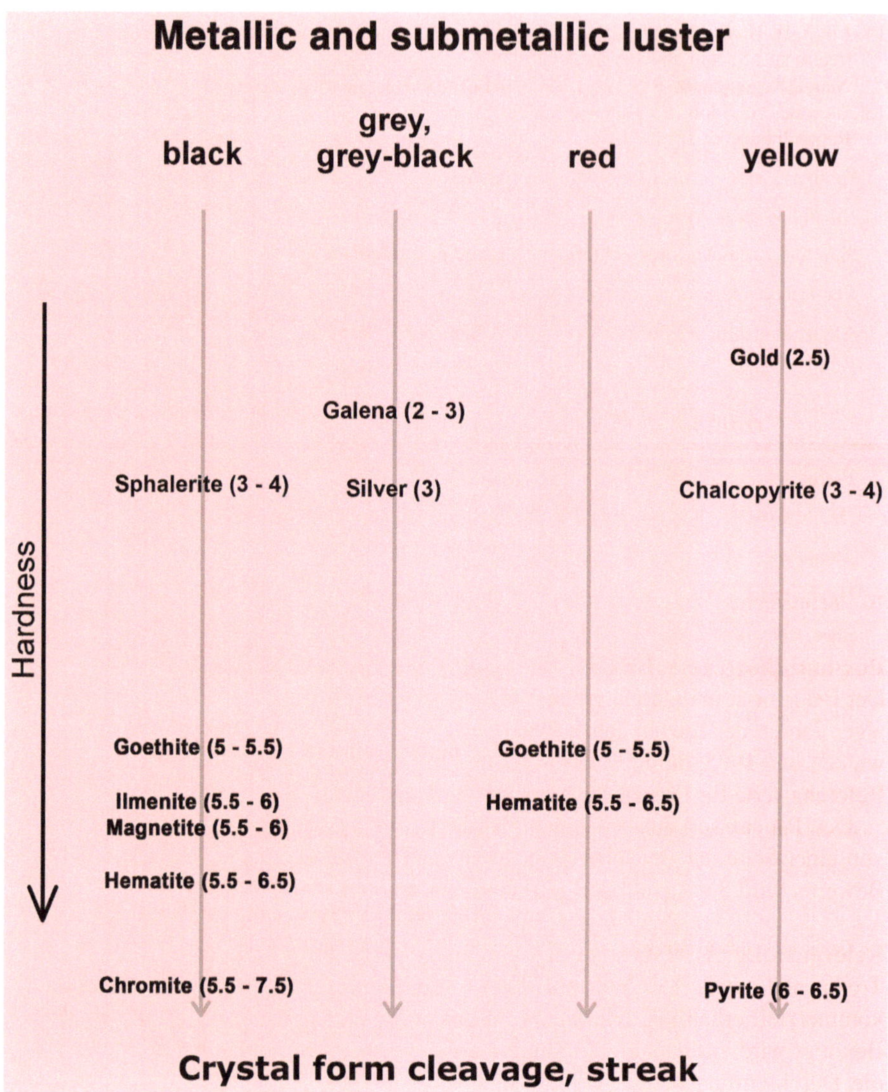

Fig. 2.5 Classification of minerals with metallic and submetallic luster based on their hardness (in brackets) and color, with examples. (After Wenk and Bulakh 2004)

2.2.2 Physical Properties

Hardness

Mineral hardness is closely related to the chemical composition and the crystalline structure of the mineral—and, in particular, to the stability of the chemical bonds between the individual atoms/ions.

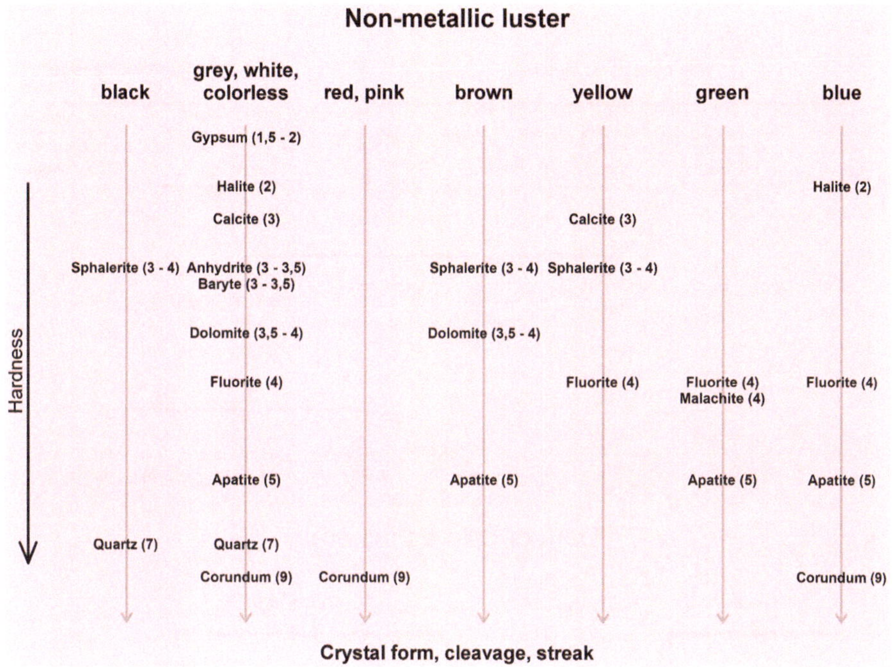

Fig. 2.6 Classification of minerals with metallic and submetallic lusters. (After Markl 2004)

The hardness of a mineral defines its resistance to scratching, and is measured using the **Mohs scale**. Each mineral can scratch the minerals above it on the scale:

1. Talc (soft)
2. Gypsum
3. Calcite
4. Fluorite
5. Apatite
6. Feldspar (Orthoclase)
7. Quartz
8. Topaz
9. Corundum
10. Diamond (hard)

Minerals with a hardness of 1 feel soapy or greasy. A fingernail has a hardness of about 2.5. A copper coin has a hardness of about 3. A steel pocket knife has a hardness of about 5.5. Minerals with a hardness of >6 can scratch glass.

Cleavage and Fracture

Cleavage is the tendency of a mineral to preferentially break along smooth planes of weakness in the crystal lattice (Fig. 2.8). Cleavage can be **perfect** (mica—in 1 direction, calcite—in 3 directions), **good** (gypsum), **indistinct/poor** (apatite, chal-

2

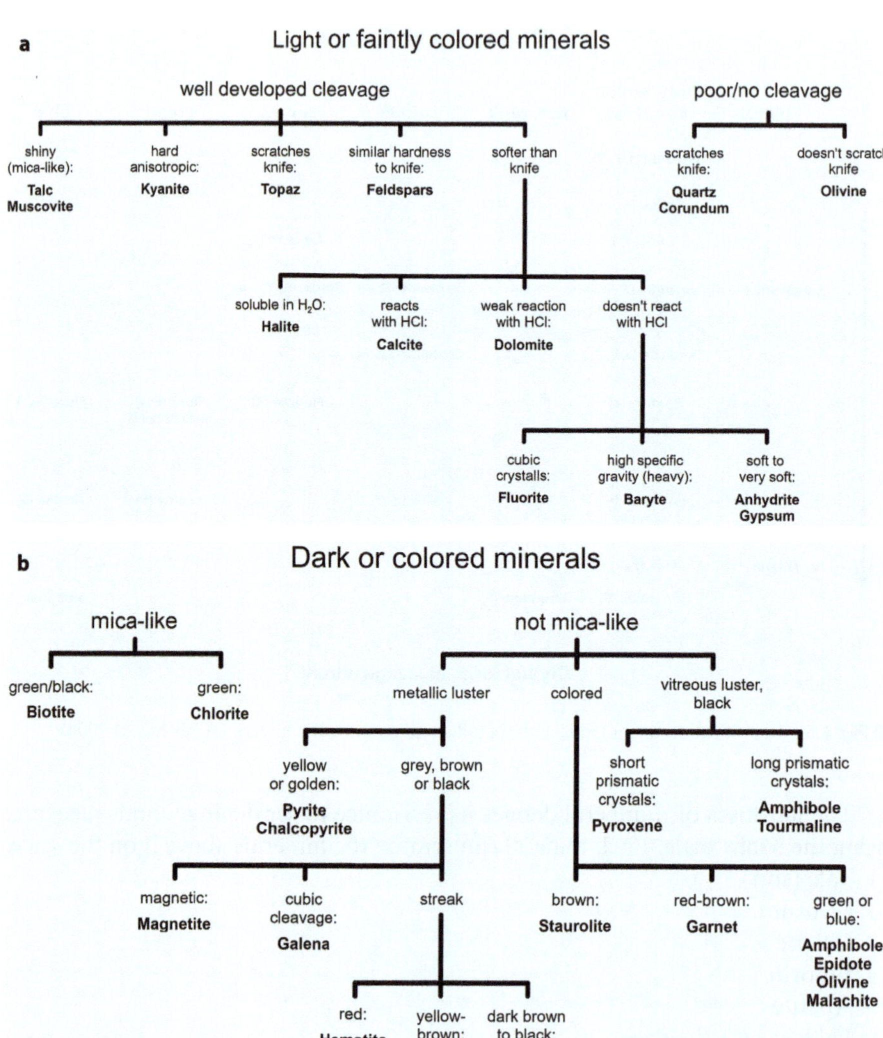

Fig. 2.7 Classification of minerals with non-metallic lusters. (After Markl 2004)

copyrite), or a mineral may have no cleavage (magnetite, garnet). Some minerals may also have more than one cleavage plane and in such cases the angle between the planes can also be of great importance for mineral identification (e.g., 90° angle for pyroxene, 124°/56° for amphibole).

Fractures occur when a mineral breaks along an irregularly-oriented surface that does not correspond to a cleavage plane (common in minerals with poor or no cleavage). Fracture can be **uneven/irregular** (e.g., pyrite, anhydrite), or **conchoidal** (i.e., smooth curved surfaces such as quartz or obsidian); these latter are also **brittle**. Subconchoidal fractures are similar to conchoidal, but not as curved.

Table 2.3 Characteristic colours of common minerals. (After Wenk and Bulakh 2004)

Minerals	Gemstone	Color
Fluorite		Violet, yellow, green, red
Halite		Mostly colorless, white, blue, yellow
Topaz		Blue, yellow
Corundum	Ruby	Red
	Sapphire	Blue
Garnet	Spessartine	Yellow orange
	Almandine	Dark red
Beryl	Emerald	Dark green
	Aquamarine	Blue-green
	Morganite	Pink to purple
Kyanite		Blue
Topaz	Imperial topaz	Golden
Tourmaline	Rubellite	Pink to red
Quartz	Amethyst	Purple
	Citrine	Yellow
	Rose quartz	Pink
	Smoky quartz	Brown
	Tiger eye	Golden yellow
Olivine	Peridote	Green
Turquoise		Blue

In metals, the fracture may be **hackly**, while fibrous minerals may have a **splintery** fracture (e.g., kyanite).

2.2.3 Other Properties

Some minerals have additional properties that can be important in distinguishing them. Examples, include:

Feel – Talc and serpentine feel slippery or “soapy” when rubbed between the fingers.

Taste – Certain water-soluble minerals have a distinct taste (e.g., halite—salty).

Tenacity – the reaction of a mineral to stress. Minerals can be brittle, sectile (can be cut with a knife), malleable (can be flattened into a sheet), ductile (can be stretched into a wire), or flexible.

Magnetism – Some minerals (e.g., magnetite) are magnetic.

Density – Some minerals have a high specific gravity due to their chemical composition, and thus feel heavy in the hand (e.g., galena).

2

Table 2.4 Characteristic streaks of common minerals. (After Wenk and Bulakh 2004)

Streak	Mineral
Metallic streak	
Gold-yellow	Gold
Silver-white	Silver
Copper red	Copper
Non-metallic streak	
Black	Graphite, ilmenite, magnetite
Greenish-black	Chalcopyrite, pyrite
Brownish-black	Pyrite, marcasite
Gray-black	Galena, marcasite, arsenopyrite (dark)
Gray	Graphite, stibnite, molybdenite, (blue to green)
Brown	Sphalerite (pale to colorless), rutile (pale)
Brownish-red	Hematite, manganite
Brownish-yellow	Goethite
Red	Hematite (dark)
Green	Malachite (pale)
Blue	Azurite (pale), lazurite

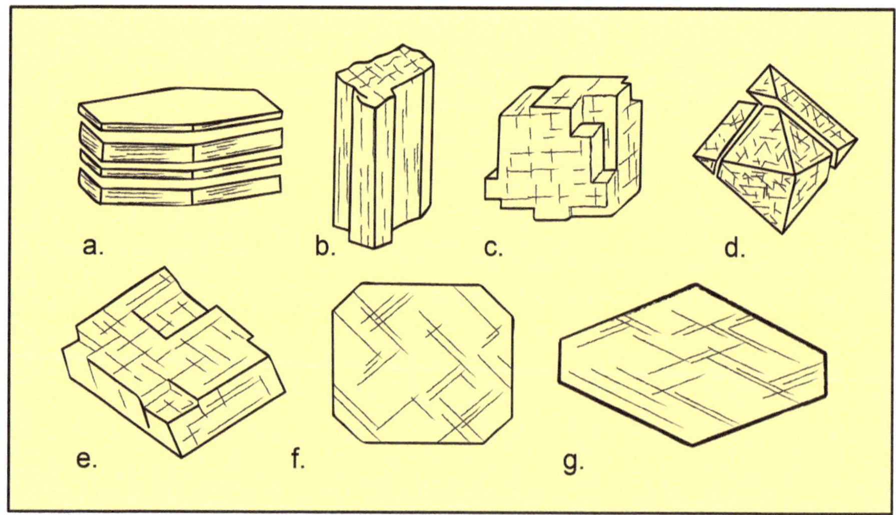

Fig. 2.8 Examples of cleavage in minerals. **a** Single cleavage causing mineral to break into flakes (mica). **b** two dominant cleavages resulting in prismatic or fibrous fragments (amphibole). **c** Three cleavages at 90° producing cubic fragments (halite) **d** octahedral cleavage (fluorite). **e** Symmetrical trigonal cleavage (calcite). The angle between the cleavage faces can also be used to identify the minerals, e.g., **f** pyroxene, and **g** amphibole. (After Wenk and Bulakh 2004)

2.3 Selected Minerals

2.3.1 Native Elements

- **Gold, Au—Cubic**

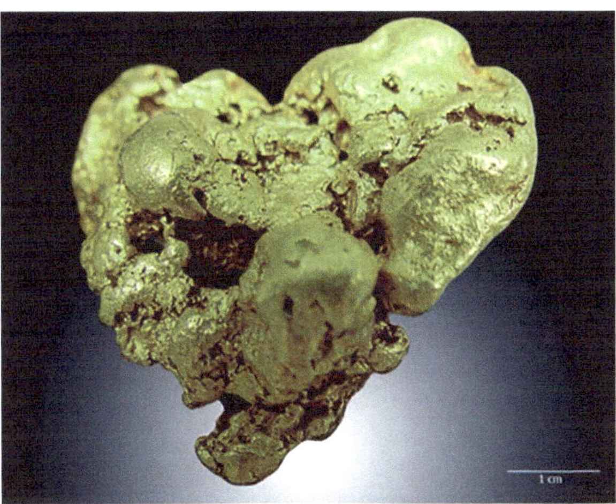

Gold (Nugget)

Habit – grains, irregular rounded masses (nuggets), dendritic forms; rarely crystals

Hardness – 2.5–3

Cleavage – none

Fracture – hackly

Color/Transparency – characteristic golden yellow; opaque

Streak – gold-yellow

Luster – metallic

Distinguishing features – color, low hardness, gold can be confused with pyrite or chalcopyrite, but differs greatly in hardness, color, ductility and malleability

Occurrence – hydrothermal veins, often associated with quartz, also concentrated in alluvial sediments (due to density)

■ Silver, Ag—Cubic

2

Silver (Photo J. Burrow)

Habit – wire-, hair- or moss-shaped to dendritic aggregates, crystals are rare

Hardness – 2.5–3

Cleavage – none

Fracture – hackly

Color/Transparency – silver-white, tarnishes quickly and is then usually yellow to brown/black due to the silver sulfide coating; opaque

Streak – silver-white-yellowish

Luster – metallic

Distinguishing features – color, tarnish, ductility and malleability

Occurrence – hydrothermal veins

2.3.2 Sulfides

- **Galena, PbS—Cubic**

1 cm

Galena

Habit – often cubes or octahedra, also blocky, massive or granular aggregates

Hardness – 2.5–3

Cleavage – cubic, perfect

Color/Transparency – lead gray, occasionally matt tarnish; opaque

Streak – gray-black

Luster – metallic

Distinguishing features – color, metallic luster, perfect cubic cleavage, high density

Occurrence – the most important and common Pb ore mineral, often found in hydrothermal deposits, pegmatites and sedimentary rocks

▪ Sphalerite, ZnS—Cubic

Sphalerite

Habit – often tetrahedral or rhombododecahedral, also granular or botryoidal aggregates

Hardness – 3.5–4

Cleavage – perfect

Fracture – conchoidal to uneven

Color/Transparency – often yellow, red-brown, green or black; transparent to translucent, never completely opaque

Streak – dark brown to yellowish

Luster – greasy, adamantine especially on cleavage surfaces

Distinguishing features – cleavage, luster, and color (although variable, it is usually yellow to dark brown)

Occurrence – most important Zn ore metal, often found in hydrothermal deposits and limestones

Chalcopyrite, $CuFeS_2$—Tetragonal

Chalcopyrite

Habit – crystals often tetrahedra, also massive or rarely botryoidal aggregates

Twinning – penetration twins, sometimes contact twinning (spinel type)

Hardness – 3.5–4

Cleavage – poor

Fracture – conchoidal, uneven

Color/Transparency – brass yellow, often with greenish tarnish; opaque

Streak – green-black

Luster – metallic

Distinguishing features – can be distinguished from pyrite by color and hardness, from gold by hardness and brittle nature

Occurrence – important Cu ore metal, often found in hydrothermal deposits and igneous and metamorphic rocks

Pyrite, Fe_2S—Cubic

Pyrite

Habit – often cubes (with striations), pyritohedron and octahedron, or combinations; also massive, globular, or radiating aggregates

Twinning – penetration and contact twins

Hardness – 6–6.5

Cleavage – indistinct

Fracture – uneven to conchoidal

Color/Transparency – brass yellow; opaque

Streak – green-black

Luster – metallic

Distinguishing features – distinguished from gold by hardness, habit and brittle nature, and from chalcopyrite by color and hardness

Occurrence – often found in hydrothermal veins, sedimentary (black shales) and metamorphic (contact zones) rocks. Often replaces fossils

2.3.3 Halides

- **Halite, NaCl—Cubic**

Halite (Photo J. Burow)

Habit – mainly cubes, but also granular or sometimes fibrous aggregates. Occasionally, cubic pseudomorphs of clay after halite are found

Hardness – 2.5

Cleavage – cubic, perfect

Fracture – conchoidal

Color/Transparency – colorless or white, sometimes yellow, gray, brown-black, red and blue; transparent to translucent

Streak – white

Luster – vitreous

Distinguishing features – soluble in water, perfect cubic cleavage, salty taste

Occurrence – a common evaporite mineral, often interbedded with sylvine, anhydrite or gypsum. Halite crusts are also found in deserts or along the margins of salt lakes.

▪ Fluorite, CaF_2—Cubic

Fluorite

Habit – mainly cubes, sometimes octahedral or intergrown/twinned as tetrakis hexahedron or hexakis octahedron, also as colored banded aggregates. May show zoning

Hardness – 4 (standard mineral on the Mohs hardness scale)

Cleavage – cubic/octahedral, perfect

Fracture – conchoidal, brittle

Color/Transparency – color is variable, often yellow, green, violet, blue, purple, sometimes colorless, pink, red, and black; translucent to transparent. Many fluorites fluoresce weakly in UV light, due to the presence of rare earth elements in the crystal structure

Streak – white

Luster – vitreous

Distinguishing features – cubic habit, cleavage, fluorescence

Occurrence – common in magmatites and ore deposits, also in hydrothermal veins

2.3.4 Oxides and Hydroxides

- **Spinel Group—Spinel, $MgAl_2O_4$—Cubic**

Spinel

Habit – often as octahedral crystals, also massive aggregates

Hardness – 7.5–8

Cleavage – none

Fracture – conchoidal, brittle

Color/Transparency – variable, often red, but also blue, green, brown, black or colourless; transparent to opaque

Streak – white, but also gray or brown

Luster – vitreous

Distinguishing features – habit, twin shape (spinel twins), hardness. Substitution of Mg atoms by Fe, Zn, and Mn atoms results in differences in color and physical properties

Occurrences – found in igneous (e.g., gabbro) and metamorphic (e.g., carbonates in areas of contact metamorphism, slate) rocks. The deep red colored spinel is a valuable gemstone

▪ Magnetite, Fe_3O_4—Cubic

Magnetite

Habit – often as octahedral crystals, less commonly dodecahedral, also as granular/massive ore

Hardness – 5.5–6.5

Cleavage – none

Fracture – subconchoidal to uneven, brittle

Color/Transparency – black; opaque

Streak – black

Luster – metallic to submetallic

Distinguishing features – strongly magnetic, color, streak, hardness

Occurences – found in igneous and metamorphic rocks and in areas of hydrothermal replacement; as a differentiation product of basic magmatic rocks (e.g., diabase) it forms important iron ore deposits

■ Corundum, Al_2O_3—Trigonal

Corundum (Photo R. Schumacher)

Habit – often columnar or flat, tabular, barrel-shaped crystals. Also bipyramidal hexagons wider in the center and tapering at the ends

Hardness – 9 (standard mineral on the Mohs hardness scale)

Cleavage – none

Fracture – uneven to conchoidal

Color/Transparency – two main types: blue (sapphire, contains Fe and Ti) and red (ruby, contains Cr). Also yellow, brown, green, or colorless; translucent to transparent/opaque

Streak – white

Luster – vitreous, adamantine

Distinguishing features – hardness, density, habit

Occurrence – found in magmatic (e.g., nepheline syenite, nepheline syenite pegmatite) and metamorphic (e.g., marble, hornfels, gneiss) rocks. As placer deposits in some alluvial sands and gravels

▪ Hematite, Fe_2O_3—Trigonal

Hematite

Habit – tabular or rhombohedral, sometimes thin plates. Often massive, mammilated, botryoidal, reniform, or radial aggregates

Hardness – 5–6

Cleavage – none

Fracture – uneven, brittle

Color/Transparency – red, steel gray to black, sometimes banded or iridescent; opaque

Streak – red to red-brown

Luster – metallic, dull

Distinguishing features – streak, hardness, habit

Occurrence – an important iron ore found in banded iron formations, as well as in hydrothermal veins and sedimentary (e.g., ooliths), igneous and metamorphic rocks

Ilmenite, $FeTiO_3$—Trigonal

Ilmenite

Habit – tabular or pyramidal crystals, including granular or massive aggregates

Hardness – 5–6

Cleavage – none, occasional parting

Fracture – conchoidal, brittle

Color/Transparency – black; opaque

Streak – black to reddish brown

Luster – metallic to submetallic

Distinguishing features – streak, weakly magnetic

Occurrence – often found in magmatic rocks (e.g., gabbro, basalt), in quartz veins and some gneisses. Also as alluvial sands. An important titanium ore.

▪ Goethite, FeO(OH)—Orthorhombic

Goethite

Habit – acicular/needle-shaped or fibrous radiating crystals, sometimes botryoidal or reniform aggregates.

Hardness – 5–5.5

Cleavage – perfect

Fracture – uneven, brittle

Color/transparency – black, brown, yellow and orange; translucent to opaque

Streak – brown, yellowish

Luster – silky, dull, adamantine (crystals)

Distinguishing features – color, streak, habit

Occurrence – secondary formation through oxidation/weathering of ferrous minerals (e.g., pyrite, magnetite). Also as iron ores (oolites) and as a precipitate (e.g., bog iron ore). Replaces other minerals (e.g., pseudomorphs after pyrite, gypsum)

2.3.5 Carbonates

- **Calcite, $CaCO_3$—Trigonal**

Calcite

Habit – often tabular, acicular, or prismatic crystals (rhombohedron, scalenohedron), also as stalagmites/stalactites, columnar, fibrous, or massive aggregates

Hardness – 3 (standard mineral on the Mohs hardness scale)

Cleavage – perfect

Fracture – conchoidal, brittle

Color/Transparency – mostly colorless, milky white, also gray, yellow, green, red, purple, blue, brown and black; transparent to translucent, sometimes opaque

Streak – white

Luster – vitreous

Distinguishing features – cleavage, hardness, strong effervescent reaction with hydrochloric acid, transparent crystals show strong double refraction

Occurrence – One of the most common minerals and predominantly sedimentary in origin (e.g., limestone, marl, tufa). Also in metamorphic (e.g., marble) or magmatic (e.g., carbonatite) rocks.

▪ Dolomite, $CaMg(CO_3)_2$—Trigonal

Dolomite

Habit – rhombohedral crystals, often with curved saddle-like faces. Also massive, granular aggregates

Hardness – 3.5–4

Cleavage – perfect

Fracture – conchoidal, brittle

Color/Transparency – often white, sometimes colorless, pink, yellowish to brownish, sometimes pink; transparent to translucent

Streak – white

Luster – vitreous, pearly

Distinguishing features – similar to calcite, powdered dolomite has an effervescent reaction with hydrochloric acid

Occurrence – often a diagenetic product (replacement of Ca by Mg), also as a gangue mineral in hydrothermal veins

■ Malachite, $Cu_2CO_3(OH)_2$—Monoclinic

Malachite

Habit – banded massive, botryoidal, reniform or fibrous radial aggregates
Hardness – 3.5–4
Cleavage – perfect
Fracture – splintery/uneven, brittle
Color/Transparency – green; opaque to transluscent
Streak – light green
Luster – vitreous, silky
Distinguishing features – color, habit, effervescent reaction with hydrochloric acid
Occurrence – an oxidation product of copper deposits

2.3.6 Sulfates

■ Barite (Baryte), $BaSO_4$—Orthorhombic

Barite (with fluorite)

Habit – tabular, sometimes prismatic or fibrous crystals. Also bladed, massive, nodular or coxcomb aggregates

Hardness – 2.5–3.5

Cleavage – perfect

Fracture – uneven

Color/Transparency – colorless, white, yellow, brown, blue, green or red; transparent to translucent

Streak – white

Luster – vitreous

Distinguishing features – density (high specific gravity), habit, hardness

Occurrence – as a gangue mineral associated with lead, copper, silver, zinc, iron or nickel ores. Replaces other minerals, but also organic materials (e.g., fossils, wood). Forms characteristic concretions (e.g., desert rose)

▪ Anhydrite, $CaSO_4$—Orthorhombic

Anhydrite

Habit – rare tabular or prismatic crystals, often fibrous. Also massive or nodular aggregates

Hardness – 3–3.5

Cleavage – perfect

Fracture – uneven

Color/transparency – colorless, white, blue, gray or red; transparent to translucent

Streak – white

Luster – vitreous, pearly

Distinguishing features – cleavage (three cleavages at 90°), hardness, density

Occurrence – found in sediments (deposits directly from warm sea water), or above salt domes. Also forms due to the dehydration of gypsum

▪ Gypsum, $CaSO_4.2H_2O$—Monoclinic

Gypsum

Habit – tabular crystals, sometimes curved, also fibrous, prismatic, bladed, massive, granular aggregates

Hardness – 2 (standard mineral on the Mohs hardness scale)

Cleavage – perfect

Fracture – uneven

Color/Transparency – colorless, white, yellow, gray, red, and brown; transparent to translucent

Streak – white

Luster – vitreous, pearly, silky

Distinguishing features – hardness, cleavage, swallowtail/fishtail twins

Occurence – common in sedimentary rocks (e.g., limestones). Also associated with salt deposits and less commonly in areas where sulfide oxidation has occurred. Forms in sandy areas (e.g., desert rose) and due to the hydration of anhydrite

2.3.7 Phosphates

- **Apatite, $Ca_5(F, Cl, OH)|(PO_4)_3$—Hexagonal**

Apatite

Habit – often columnar, tabular, hexagonal or flat crystals, also as granular, fibrous, or radiating aggregates

Hardness – 5 (standard mineral on the Mohs hardness scale)

Cleavage – indistinct

Fracture – conchoidal

Color/transparency – colorless, white, yellow, green to gray-green, brown, blue or red; transparent to translucent

Streak – white

Luster – vitreous, waxy

Distinguishing features – habit, hardness

Occurrence – found in many magmatic rocks (e.g., granite pegmatite, nepheline syenite), also in high-temperature hydrothermal veins and in regional or contact metamorphic rocks (e.g., hornfels). Apatite is a main component in bones and tooth enamel

2.3.8 Silicates

Orthosilicates

- **Olivine, $(Mg, Fe)_2SiO_4$—Orthorhombic**

Olivine

A mineral group with varying compositions, ranging from forsterite (Mg_2SiO_4) through to fayalite (Fe_2SiO_4)

Habit – mostly isolated crystals (prismatic, tabular), or as granular aggregates

Hardness – 6.5–7

Cleavage – indistinct

Fracture – conchoidal

Color/Transparency – green, sometimes yellow (in basalts) or brown to black, reddish when oxidized; transparent to translucent

Streak – white

Luster – vitreous

Distinguishing features – color (olive green), fracture. Physical properties vary depending on the proportions of Mg and Fe present

Occurrence – a range of basic/ultrabasic magmatic rocks (e.g., basalt, gabbro, peridotite). Dunite comprises almost 100% olivine. Olivine is common in some meteorites, as well as in lunar basalt

▪ Garnet group, $X_3Y_2Si_3O_{12}$ (X = Ca, Mn, Mg or Fe^{2+}; Y = Al, Cr, or Fe^{3+})—Cubic

Garnet

The garnet mineral group comprises the following end members:

- Pyrope-Almandine-Spessartine Group:
  - Pyrope $Mg_3Al_2Si_3O_{12}$;
  - Almandine $Fe_3Al_2Si_3O_{12}$;
- Spessartine $Mn_3Al_2Si_3O_{12}$
- Grossular-Uvarovite-Andradite Group:
  - Grossular $Ca_3Al_2Si_3O_{12}$;
  - Uvarovite $Ca_3Cr_2Si_3O_{12}$;
  - Andradite $Ca_3Fe_2Si_3O_{12}$.

Within (but not between) each group there is continuous atomic substitution.

Habit – often well-formed dodecahedral or trapezohedral crystals, also coarse or granular aggregates

Hardness – 6.5–7.5

Cleavage – none

Fracture – conchoidal, uneven

Color/Transparency – varies with composition, dark red, brown to black (pyrope, almandine and spessartine), green (uvarovite), brown, light green or white (grossular), yellow, brown or black (andradite); transparent to opaque

Streak – white

Luster – vitreous, resinous, adamantine

Distinguishing features – hardness, habit

Occurrence – found in metamorphic (e.g., schist, gneiss) and some magmatic (e.g., peridotite, granite, pegmatite) rocks. Also common in alluvial sands (e.g., beach, river)

Group Silicates

Kyanite (Disthene), Al_2SiO_5—Triclinic

Kyanite

Habit – long, slender, bladed crystals, also radial aggregates

Hardness – 5.5–7 (hardness is variable, 5.5 along crystal length, and 6–7 across the crystal)

Cleavage – perfect

Color/Transparency – blue, white, also gray, green or black (color is often uneven); transparent to translucent

Luster – vitreous, sometimes pearly

Distinguishing features – color, habit, cleavage, variable hardness

Occurrence – found in metamorphic (e.g., schist, gneiss) and magmatic (e.g., granite pegmatite) rocks. Also in quartz veins

2

▪ Topaz, $Al_2SiO_4(OH,F)_2$—Orthorhombic

Topaz

Habit – prismatic crystals, often striated, also columnar, massive, radiating and granular aggregates

Hardness – 8 (standard mineral on the Mohs hardness scale)

Cleavage – perfect (basal)

Fracture – subconchoidal, uneven

Color/Transparency – colorless, yellow, light blue, green and pink; transparent to opaque

Streak – colorless

Luster – vitreous

Distinguishing features – habit, hardness, cleavage

Occurrence – found in acidic magmatic rocks (e.g., granite pegmatite, rhyolite), alluvial sediments and quartz veins

Staurolite, $(Fe, Mg)_2(Al,Fe)xSi_4O_{20}(O,OH)_2$—Monoclinic, Pseudo-Orthorhombic

Staurolite

Habit – prismatic crystals, sometimes tabular, often twinned (staurolite twins), rare aggregates

Hardness – 7–7.5

Cleavage – good

Fracture – subconchoidal, uneven

Color/Transparency – reddish brown to brown-black; translucent to almost opaque

Streak – white

Luster – vitreous, resinous

Distinguishing features – color, habit (especially with cross-shaped penetration twins)

Occurrence – found in metamorphic rocks (e.g., schist, gneiss)

Group Silicates

Epidote Group

General formula is $X_2Y_3Si_3O_{12}(OH)$, where X is often Ca and Y is normally Al and Fe^{3+}, partially replaced by Mg and Fe^{2+} in some forms.

Zoisite, $Ca_2Al_2Si_2O_{12}(OH)$—Orthorhombic

Habit – prismatic crystals (often striated), also radial, fibrous or massive aggregates

Hardness – 6–6.5

Cleavage – perfect

Fracture – conchoidal to uneven

2

Color/Transparency – gray, yellow, pink, blue, light green or brown; transparent (strongly pleochroic) to translucent

Streak – white

Luster – vitreous, pearly

Distinguishing features – color, cleavage

Occurrence – found in contact (e.g. gneiss, hornfels) and regional (e.g. schist) metamorphic rocks. Also in metasomatic rocks and rarely in granites.

Clinozoisite, $Ca_2Al_3Si_3O_{12}(OH)$ and Epidote, $Ca_2(Al,Fe)_3Si_3O_{12}(OH)$—Monoclinic

Epidote (scale: 2 cm)

Habit – prismatic crystals, sometimes striated, also massive, fibrous, granular or radial aggregates

Hardness – 6–7

Cleavage – perfect

Fracture – uneven

Color/Transparency – green-gray (clinozoisite), olive-green, brownish-green and yellow-green to black (epidote); transparent to almost opaque

Streak – white

Luster – vitreous

Distinguishing features – color, habit

Occurrence – found in contact (e.g., hornfels, metamorphosed limestone) and regional (e.g., schist) metamorphic rocks. Also in magmatic rocks (e.g., basalt, granite)

Ring Silicates

- **Tourmaline Group, (Ca, Na, K)(Li, Mg, Fe^{2+}, Mn^{2+}, Al, Cr^{3+}, V^{3+}, Fe^{3+}, $Ti^{4+})_3$(Mg, Al, Fe^{3+}, V^{3+}, $Cr^{3+})_6((Oh)_{4|}(BO_3)_{3|}(Si_8O_{18}))$—Trigonal**

Quartz with tourmaline

Habit – elongate prismatic crystals, often striated, and often with a triangular cross section; also columnar or radiating (rarely massive) aggregates

Hardness – 7–7.5

Cleavage – indistinct

Fracture – conchoidal, uneven

Color/Transparency – highly variable (due to composition) but usually black/blue-black, also colorless, blue, pink, purple or green; transparent to opaque

Streak – white

Luster – vitreous

Distinguishing features – habit, striation, color, cross section, hardness

Occurrence – found in acid magmatic (e.g., granite pegmatite, rhyolite) and metamorphic (e.g. schist, gneiss, marble) rocks

▪ Chain Silicates

Pyroxene Group

The general formula is $X_2Si_2O_6$, where X is often Mg, Fe, Mn, Li, Ti, Al, Ca or Na. The most common pyroxenes are Ca-, Mg-, or Fe-silicates, with two main groups—the orthopyroxenes are orthorhombic and have little Ca, while the clinopyroxenes are monoclinic and contain either Ca or Na, Al, Fe^{3+}, or Li.

▪ Orthopyroxene—Orthorhombic

Hypersthene

- Enstatite, $MgSiO_3$
- Hypersthene, $(Mg,Fe)SiO_3$

Habit – prismatic, columnar crystals, often as fibrous, or radiating aggregates

Hardness – 5–6

Cleavage – good

Fracture – uneven, brittle

Color/Transparency – gray, pale green, yellow–brown, greenish-brown, brown or green-black; transparent to opaque

Streak – grey-white, light brown

Luster – vitreous

Distinguishing features – cleavage (2 faces at 90°), color

Occurrence – found in magmatic rocks (e.g. gabbro, pyroxenite, andesite), also in some meteorites

- **Clinopyroxene—Monoclinic**
  - Diopside-Hedenbergite Series, $Ca(Mg,Fe)Si_2O_6$
  - **Augite, $(Ca,Mg,Fe,Ti,Al)(Al,Si)_2O_6$**

Habit – prismatic, tabular, columnar crystals with a rectangular or octagonal cross section, also granular, massive, or fibrous aggregates

Hardness – 5.5–6.5

Cleavage – good

Fracture – uneven, brittle

Color/Transparency – green, gray, white, yellow, light blue (diopside), dark green to black (augite), translucent to opaque

Streak – white (diopside), gray-green (augite)

Luster – vitreous, dull

Distinguishing features – color, habit, cleavage (2 faces at 90°)

Occurrence – augite is found in magmatic rocks (e.g., basalt, gabbro, pyroxenite) while diopside is present in metamorphic rocks (e.g., hornfels, skarn deposits)

Amphibole Group

- **Hornblende, $(Na, K)_{0-1}$ $(Ca,Na)_2(Mg,Fe,Al)_5(Si,Al)_8O_{22}(OH)_2$—Monoclinic**

Hornblende

Habit – prismatic crystals (sometimes 6-sided cross section), also columnar, fibrous or radiating aggregates

Hardness – 5–6

2 **Cleavage** – perfect

Fracture – uneven, brittle

Color/Transparency – light to dark green, black, dark brown; translucent to almost opaque

Streak – colorless

Luster – vitreous, submetallic

Distinguishing features – color, habit, cleavage (2 faces at 120°)

Occurrence – found in many magmatic rocks (e.g., granodiorite, diorite, syenite, gabbro, and the associated extrusive volcanics), but also in metamorphic rocks (e.g., hornblende schist, amphibolite)

Sheet Silicates

- **Talc, $Mg_3Si_4O_{10}(OH)_2$—Monoclinic**

Talc

Habit – rare crystals, often granular, foliated, radiating or fibrous aggregates (often forms pseudomorphs of other minerals)

Hardness – 1 (standard mineral on the Mohs hardness scale)

Cleavage – perfect

Fracture – uneven

Color/Transparency – white, gray, light green, yellow or blue; translucent to opaque

Streak – white to pale green

Luster – pearly, greasy, waxy

Distinguishing features – hardness, greasy/soapy feel, color

Occurrence – often an alteration product of olivine, pyroxene, and amphibole. Within faults in Mg-rich rocks. Also in metamorphic rocks (e.g. schist, altered dolomite)

Mica Group

There are two main mica groups—one rich in Fe and Mg (biotite) and one rich in Al (muscovite).

Muscovite, $KAl_2(AlSi_3O_{10})(OH,F)_2$—Monoclinic, Pseudo-Hexagonal

Muscovite

Habit – tabular, foliated crystals (sometimes with a hexagonal outline), also scaly, or flaky aggregates

Hardness – 2–2.5

Cleavage – perfect (individual leaves are flexible and elastic)

Fracture – uneven

Color/Transparency – colorless, white, yellow, gray, green or brown; transparent to translucent

Streak – colorless

Luster – vitreous, pearly

Distinguishing features – cleavage, color

Occurrence – commonly found in magmatic (e.g., granite, pegmatite) and metamorphic (e.g., schist, gneiss) rocks. Also as an alteration product (sericite)

2

(Phlogopite) Biotite Series

- Phlogopite, $KMg_3AlSi_3O_{10}(OH,F)_2$monoclinic
- Biotite, $K(Mg,Fe)_3AlSi_3O_{10}(OH,F)_2$monoclinic

Biotite

Habit – tabular flakes (sometimes hexagonal outline), also foliated, or scaly aggregates

Hardness – 2.5–3

Cleavage – perfect

Fracture – uneven

Color/Transparency – brown, reddish brown, yellow brown, yellow or green (phlogopite), black, dark brown or dark green (biotite); transparent to translucent

Streak – colorless

Luster – vitreous, pearly, (submetallic on cleavage faces)

Distinguishing features – cleavage, habit

Occurrence – phlogopite is found in Mg-rich magmatic rocks and metamorphosed limestones and dolomites. Biotite occurs in granites, syenites, diorites, and their volcanic equivalents and in some metamorphic rocks (e.g., schist, gneiss)

■ Chlorite Group, $(Mg,Fe,Al)_6(Si,Al)_4O_{10}(OH)_8$—Monoclinic

A collective name for a group of minerals with similar compositions.

Chlorite

Habit – pseudohexagonal tabular crystals, also foliated, scaly, or platy aggregates

Hardness – 2–3

Cleavage – perfect (individual scales are flexible, but not elastic)

Color/Transparency – green, black, more rarely yellow, red, brown; transparent to translucent

Streak – white

Luster – vitreous, pearly, dull

Distinguishing features – color, cleavage, slightly greasy feel

Occurrence – in metamorphic rocks (e.g., slate, schist) or as an alteration product of other minerals (e.g., pyroxene, amphibole, mica) in magmatic rocks. Also as amygdules in vesicular volcanic rocks or as detrital grains in sediments

Framework Silicates

■ Quartz, SiO_2—Trigonal

Quartz: **a** Quartz crystal aggregate (Rock crystal), **b** Smoky quartz

Habit – crystals are mostly prismatic, varying in shape and size, the prisms are mostly hexagonal (often with horizontal striations). Also drusy, granular, bladed, globular and massive aggregates

Hardness – 7 (standard mineral on the Mohs hardness scale)

Cleavage – none

Fracture – conchoidal

Color/Transparency – pure quartz is colorless; the main colored varieties are listed below; transparent to opaque

Luster – vitreous

Varieties of quartz – there are many varieties, distinguished by color, transparency, and other characteristics: rock crystal (colorless, transparent), smoky quartz (smoky brown, transparent to translucent), citrine (lemon yellow, transparent to translucent), amethyst (violet to purple, translucent) and rose quartz (pink to cloudy pink)

Distinguishing features – habit, fracture, luster, hardness

Occurrence – quartz is a common mineral in many magmatic and metamorphic rocks, especially granite and gneiss, but also as detrital grains in clastic sediments. Also a common gangue mineral

▪ Chalcedony, SiO_2—Trigonal

Habit – microcrystalline variety of quartz, occurs as mammilary, stalactitic, massive, radial or bulbous aggregates

Hardness – 6.5–7

Cleavage – none

Fracture – conchoidal, brittle

Color/Transparency – white, blue, red, green, pink, brown, black, colorless and often banded; transparent to translucent

Luster vitreous, waxy

Varieties of chalcedony – the varieties differ in color, transparency, and other characteristics: carnelian (pink, red), agate (finely banded), moss agate (milky white with dendritic growths of manganese oxide), onyx (black and white bands), jasper (opaque, intensely colored chalcedony, usually brown, red, yellow, or green), flint and chert (grey to black, opaque)

Distinguishing features – habit, fracture, hardness

Occurrence – found as cavity linings, in geodes and veins in a variety of rocks. It also forms pseudomorphs, for example, coral or wood (petrified wood)

▪ Opal, SiO_2 n H_2O—Amorphous

Habit – massive, botryoidal, stalactitic, nodular and rounded forms, also as veins

Hardness – 5.5–6.5

Cleavage – none

Fracture – conchoidal

Color/Transparency – colorless, white, gray, red, brown, blue, green, black, and banded; transparent to milky translucent and opaque. Precious opals show a vivid play of colors (opalescence)

Luster – vitreous, waxy

Distinguishing features – habit, opalescence, density (less dense than chalcedony)

Occurrence – found in veins in many rock types, but especially next to geysers or hot springs. Opal also forms the skeleton of many organisms (e.g., radiolaria, diatoms, sponges); their deposits can form opal-rich sediments (e.g. diatomite)

- **Feldspar Group**

Feldspars are the most common minerals in the Earth's crust (>60%), particularly in metamorphic and magmatic rocks. Their composition is XAl(Si, Al)Si_2O_8,
 where X = K, Na, Ca, Br, and Sr. As a result of these differences in composition, the crystal forms and properties show a degree of variability.

- **Alkali feldspars (K, Na)$AlSi_3O_8$: Sanidine, monoclinic; Orthoclase, monoclinic; Microcline, triclinic**

Orthoclase

Sanidine

Habit – Sanidine crystals are often tabular, sometimes prismatic—Twinning is common. Orthoclase forms both tabular and prismatic crystals (sometimes striated). Carlsbad twinning is common. Microcline is very similar to orthoclase, individual crystals can be large. All three can also occur as crystal aggregates.

Hardness – 6.0 (standard mineral on the Mohs hardness scale)

Cleavage – perfect (cleavage angle is c. 90°)

Fracture – conchoidal to uneven

Color/Transparency – Sanidine is colorless, yellow, gray or brown; translucent to transparent. Orthoclase is colorless, white, pink, blue, green, or grey; transparent to opaque. Microcline is similar to orthoclase; may also be adularescent.

Streak – white

Luster – vitreous, pearly

Distinguishing features – color, habit, cleavage, and hardness distinguish orthoclase and microcline from other minerals, but it is difficult to distinguish them from each other. Sanidine can be distinguished based on its transparency, tabular habit and occurrence. Carlsbad twins occur in both orthoclase and sanidine

Occurrence – sanidine, orthoclase and microcline form the alkali/potassium feldspar group. They are polymorphs, with almost identical physical properties. Orthoclase is the most common alkali feldspar in both metamorphic and magmatic (e.g., pegmatite) rocks. Microcline is common in metamorphic and magmatic (e.g., granites, granite pegmatite) rocks and hydrothermal veins. Sanidine is the high-temperature form of $KAlSi_3O_8$ and is often found as phenocrysts in relatively young volcanic rocks (rhyolite, trachyte) and their tuffs. Detrital grains of all three are also are found in sedimentary rocks

■ Plagioclase Series $NaAlSi_3O_8$-$CaAl_2Si_2O_8$—Plagioclase, Triclinic

Plagioclase

Habit – tabular or prismatic crystals (often striated), often grouped or twinned, also massive, grainy, columnar aggregates

Hardness – 6.0–6.5

Cleavage – good (cleavage angle is c. 90°)

Fracture – conchoidal, uneven

Color/Transparency – white, colorless, pink, green, blue, brown and black; transparent to translucent

2

Luster – vitreous, pearly on cleavage faces

Distinguishing features – habit, twin lamination (albite twinning) is pronounced on cleavage surfaces. Labradorite often shows marked blue/green colors on cleavage faces.

Occurrence – the plagioclase series comprises six different minerals (albite, oligoclase, andesine, labradorite, bytownite, anorthite) with varying amounts of Na and Ca. It is extremely difficult to distinguish them in the field. Plagioclase is a common mineral in magmatic (acid rocks are more Na rich, while basic rocks are more Ca rich) and metamorphic rocks, but also occurs detritally in sedimentary rocks

Feldspathoid/Foid Group

Chemically related to feldspars, but with a lower SiO_2 content.

- **Leucite, $KAlSi_2O_6$—Normally Tetragonal (pseudo-cubic); Cubic >625 °C**

Leucite

Habit – pseudo-isometric crystals (trapezohedrons), also granular aggregates

Hardness – 5.5–6

Cleavage – none

Fracture – conchoidal, uneven

Color/Transparency – white, gray; transparent to opaque

Streak – white

Luster – vitreous, dull

Distinguishing features – habit

Occurrence – leucite is unstable under high pressure conditions and is never found in combination with quartz. Therefore, the occurrence is limited. Typically present in K-rich, low-Si volcanic rocks (e.g., trachyte). Sometimes altered to pseudoleucite, a mixture of orthoclase and nepheline

- **Nepheline, $NaAlSiO_4$—Hexagonal**

Nepheline

Habit – tabular or prismatic crystals (usually 6-sided outline), also granular, coarse aggregates

Hardness – 5.5–6

Cleavage – poor

Fracture – subconchoidal, uneven

Color/Transparency – white, gray, colorless, brown, green; transparent to opaque

Streak – white

Luster – greasy, vitreous

Distinguishing features – luster, greasy feel

Occurrence – found in SiO_2-poor intrusive and extrusive magmatic rocks (e.g., nepheline syenite, phonolite)

Magmatic Rocks

Contents

T. McCann, *Pocket Guide Geology in the Field*,
https://doi.org/10.1007/978-3-662-63082-2_3

3

Magmatic rocks are often exposed at the Earth's surface, either because they cooled at, or close to, the surface (e.g., volcanoes) or because they have been exhumed as a result of tectonics and erosion following solidification at depth (e.g., exhumation of large intrusive rock bodies). Understanding the formation and development of a magmatic province, requires a variety of field and laboratory investigations, with more detailed characterization of a magmatic body requiring chemical and isotopic analysis. Magmatic rocks can be divided into two groups: **intrusive rocks**, which cool and solidify from a magma within the Earth and **extrusive rocks**, which cool and solidify at the Earth's surface. Hypabyssal (subvolcanic) intrusions are emplaced at shallow depths (<2 km).

3.1 Intrusive Magmatic Rocks—Types of Intrusive Bodies

Intrusive magmatic rocks are formed by the cooling and solidification of magma deep within the Earth's crust (▫ Fig. 3.1, ▫ Table 3.1). Individual magmatic bodies range in size from meters up to several kilometers in diameter.

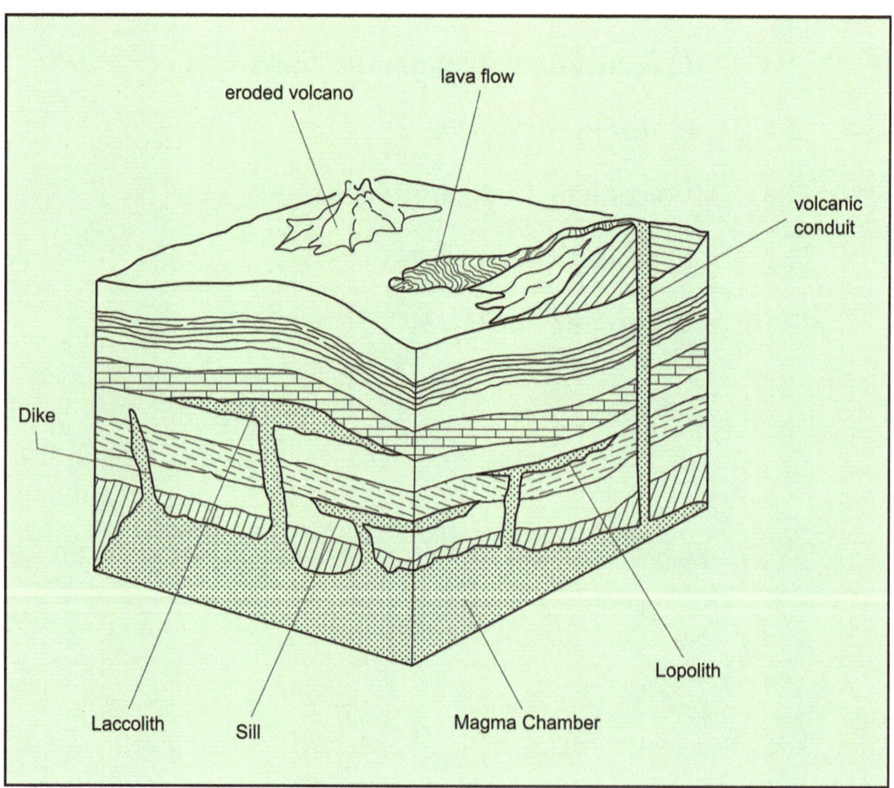

▫ **Fig. 3.1** Variety of intrusions showing their possible relationships with a sub-volcanic magma chamber. (After Thorpe and Brown 1985)

Table 3.1 Intrusive bodies and their dimensions. (After McCann and Valdivia Manchego 2015)

	Thickness	Width/length/area	Composition
Sills	Several m to hundreds of m	Up to 10 km wide	Mainly mafic
Laccoliths	Max. approx. 1000 m	1–8 km	Mainly Si-rich rocks
Lopoliths	Several m to several km	Several tens to hundreds of km in diameter	Often asymmetrical and layered; mainly mafic to ultramafic rocks
Dikes	<1 m to several hundred m	Up to several tens of km	Si-rich to mafic to ultramafic
Batholiths	–	Tens of km wide; 100 to thousands km^2	Mainly Si-rich rocks
Stocks	–	Several km wide; maximum 100 km^2	Mainly Si-rich rocks
Volcanic plugs (vent fillings)	Several hundred meters to 1 km	–	Variable, depending on chemistry of the volcano

Dikes and Veins

Vertical dike (Tenerife)

Planar, discordant intrusive bodies that cut through bedding and foliation in the surrounding host rock (Fig. 3.1). Thicknesses range from <1 m to several hundreds of meters. They can form groups (**swarms**) which can be parallel or radial (e.g., on the flanks of a volcano). Small dikes (mm to cm scale, e.g., aplites) are also called veins.

3

Diabase vein in granite

Granite vein in diorite

Sills

Concordant, layered intrusive bodies that are oriented more or less parallel to the stratification or foliation within the host rock (Fig. 3.1). Their thickness can range from meters to several hundred meters and they can extend over areas of tens to hundreds of square meters. They are generally formed by low viscosity magmas. Sills occur individually or in groups.

▪ Laccoliths

Laccolith (dacite), Maiden Creek Sill, USA. (Sill about 3 m thick; Photo C. Breitkreuz)

Mushroom-like concordant intrusions, which form mainly at depths of around 3 km below the Earth's surface (◘ Fig. 3.1). Laccoliths have a thickness of up to 1 km and a diameter of 1–8 km. In general, laccoliths are formed from silica-rich magmas (higher viscosity than mafic magmas) and, therefore, there is little lateral spreading.

▪ Lopoliths

Concordant, lenticular, saucer- or funnel-shaped intrusions with thicknesses ranging from meters to kilometers and diameters that can reach several tens to several hundreds of kilometers. They often form layered intrusions (i.e., bodies with layers which differ in terms of composition or texture).

▪ Batholiths and Stocks

Large, coarsely-crystalline, plutonic bodies, which often form elongated intrusive belts (50–150 km wide and 500–1500 km long). Stocks are smaller structures with a maximum surface area of 100 km^2. Batholiths usually consist of a large number of overlapping, smaller intrusive bodies or plutons (each 5–50 km in diameter). Batholiths and stocks are often silica-rich rocks.

3

▪ Volcanic Plugs

Volcanic plugs, Agathla Peak, USA

The erosion of volcanic bodies (e.g., volcanic vents) can expose round to oval structures in the outcrop (volcanic plugs/necks). These have a diameter of around 10^2–10^3 m and include lavas and pyroclastic material. Internally they often show brecciation due to the passage of volcanic gases and hydrothermal solutions.

3.2 Volcanoes

Volcanic eruptions are among the most dramatic and visible magmatic processes (◘ Fig. 3.2). Most volcanic activity (about 90%) takes place along tectonic plate boundaries (e.g., Ring of Fire around the Pacific).

Volcanic cones (**stratocones**, although other cones also exist, e.g., cinder cones etc.) typically have a central crater, which overlies the **vent** (although the formation of several secondary vents is possible), in which the lava moves from the underground **magma chamber** (or possibly several magma chambers) toward the Earth's surface. However, eruptions along fissures (e.g., flood basalts) are of much greater importance in the Earth's history because of their high volume.

Volcanoes occur in a variety of forms. The particular shape is largely related to the composition of the magma, which also determines the type of eruption. For example, basic lavas (e.g., basalt) have a low viscosity and therefore flow easily, so that they form wide **shield volcanoes**. Magmas with a higher proportion of SiO_2 (acid lavas, e.g., rhyolite) are more viscous and explosive. They form **layered volcanoes**, which are built up of lava and fragmented material. Magmas rich in silica can be so viscous that the impinging magma can cause the volcano to bulge and deform (e.g., Mount St Helens, USA).

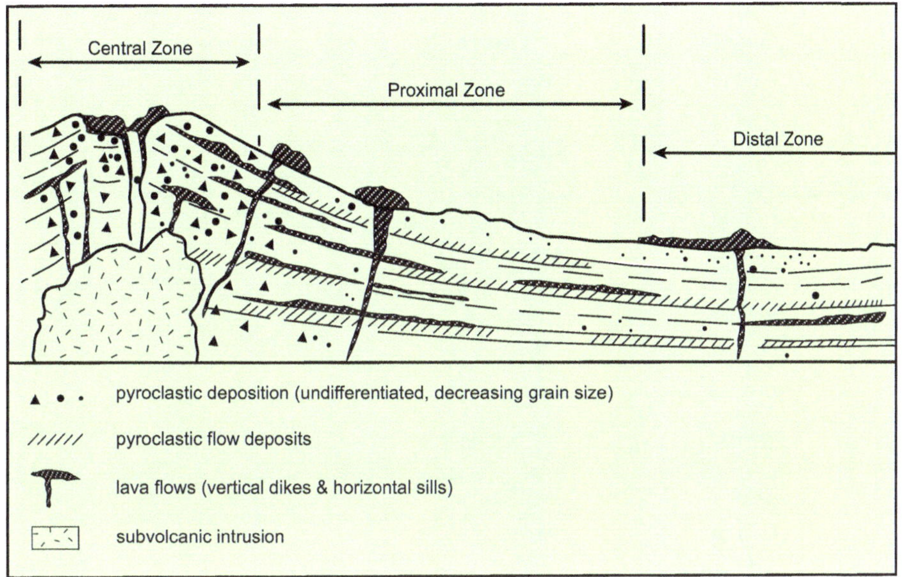

Fig. 3.2 Schematic profile showing the facies variations in a volcano. (After Thorpe and Brown 1985)

3.3 Volcanic Eruption Types

Magmatic eruptions can be divided into the following main types (Fig. 3.3; note that transitions between the various types occur):

- **Icelandic eruptions**—least explosive; fault or fissure-related eruptions, mostly produced by low-viscosity magmas.
- **Hawaiian eruptions**—are closely related to Icelandic eruptions and are formed by basic (basaltic) magmas. Eruptions are sporadic. Eruption phases of gas-poor lavas alternate with short phases of gas-rich eruptions (e.g., fire fountains or curtains of fire).
- **Strombolian eruptions**—magmas have a higher viscosity than those of Hawaiian eruptions. Explosive eruptions are produced in an open vent due to the bursting of large gas bubbles which form as a result of reduced pressures in the upper part of the magma column.
- **Vulcanian eruptions**—similar to Strombolian eruptions, but tend to be more explosive. They are often associated with highly viscous magmas of andesitic composition.
- **Peléan eruptions**—are associated with avalanches of hot volcanic ash (pyroclastic flows). Lava domes also form, which may subsequently collapse.
- **Plinian (Vesuvian) eruptions**—are the largest and most explosive eruptions (mainly highly viscous acid to intermediate magmas). The eruptions are marked by the production of high columns of gases and volcanic debris.

3

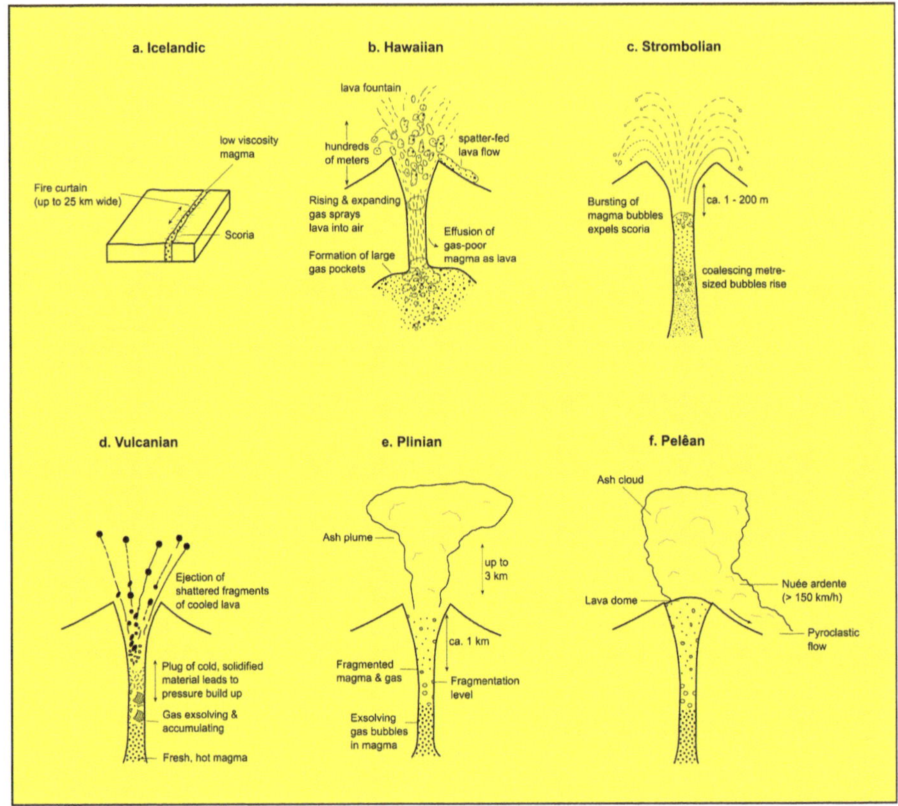

Fig. 3.3 Schematic diagram showing the main eruption types. (After Orton 1996; McCann and Valdivia Manchego 2014)

In addition to the various eruption types, there are also **phreatomagmatic eruptions**, where the erupting magmas come into contact with groundwater, seawater, or ice (e.g., beneath a glacier). These are usually explosive and produce a lot of water vapor and pyroclastic material.

3.4 Volcanic Deposits

There are two types of volcanic deposits: lavas and pyroclastics. Both usually form during a volcanic eruption. Lavas are formed when magma emerges from the volcano and solidifies on contact with the atmosphere or water. Four main lava types are recognised, depending on their form and surface morphology.

3.4.1 Lavas

■ Aa lava

Aa lava (Hawaii; scale 30 cm)

Lavas with a rough, uneven or rubbly (brecciated) surface, containing entrained fragments of older crust. Lava flow was rapid.

■ Pahoehoe Lava

Pahoehoe lava (Hawaii; scale 5.5 cm)

Lavas with smooth or rope-like crusts, which often have a basic composition. They originate from hotter, less viscous lavas than the Aa type.

3

▪ Blocky Lava

Blocky lava and lava ball (Hawaii; picture width approx. 4 m)

Similar to aa lava, with irregular surfaces consisting of large standing blocks (dm's-m's) and typically comprise andesitic, dacitic or rhyolitic material.

▪ Pillow Lava

Pillow basalts (scale 30 cm)

Consists of round to elliptical structures with a diameter of up to 1 m, which form when lava flows out underwater (or under ice). Compositions are usually basaltic or andesitic.

Columnar Jointing

Basalt columns, Giants Causeway, Northern Ireland (column diameter approx. 30 cm; Photo B. Murphy)

Basalt columns, Mammoth Lake, USA (scale approx. 2–4 m)

These structures occur in a variety of igneous rocks, and form as the rocks cool and contract. In basaltic lavas, sills, and dikes, five- to seven-sided columnar structures often occur. These columns form during cooling and are arranged perpendicular to the upper- and lower-cooling surfaces.

3.4.2 Pyroclastics

Pyroclastic materials (e.g., pumice, bombs) are produced from most volcano types (with the exception of Hawaiian and Icelandic) and comprise a range of materials formed due to differences in gas content and viscosity. These latter result in variations in vesicularity (i.e., gas bubbles in the magma) and crystallinity. In subaerial, submarine, or subglacial eruptions, pyroclastic material is initially ejected vertically in the form of an eruption column (◘ Fig. 3.3). The pyroclastic material (**tephra**) formed during volcanic eruptions contains three main types of fragments (◘ Fig. 2.29):

- **Pyroclastic fragments:** These are divided into fine ash (<0.063 mm), coarse ash (0.063–2 mm), lapilli (2–64 mm), and blocks (>64 mm; solid ejecta) or bombs (>64 mm; molten ejecta) according to their grain size. The lithified equivalents are fine tuff, coarse tuff, lapillstone/volcanic conglomerate, pyroclastic breccia and agglomerate, respectively;
- **Glass fragments;**
- **Single crystals.**

A variety of pyroclastic types have been recognised. **Pyroclastic fall deposits (ash deposits)** are initially transported upwards as part of an eruption plume. The fragments subsequently fall as a result of gravity and form a blanket-like deposit around the collapsing eruption column.

Pyroclastic density flows (i.e., flows, surges) can be subdivided on the basis of their density and turbulence. **Pyroclastic flow deposits (ignimbrites)** are gravitational flow deposits of volcanic fragments that spread around the volcano in the form of hot, highly concentrated mixtures of gas and fragments (◘ Fig. 3.2). Velocities close to the eruption column can be up to 1000 km/h. The pyroclastic flows, in which temperatures can be up to 1000 °C, destroy all vegetation over a radius of over 100 km. **Pyroclastic surge deposits** are also gravitational flows, but with a lower concentration (density) of fragments than pyroclastic flow deposits. They spread laterally as hot gas–solid mixtures and show a variety of sedimentary structures that indicate the direction of flow (e.g., cross-bedding, planar lamination).

A mixture of volcanic material ± sediment and water (ice, snow) can produce debris flows (**lahars**). These are very dangerous and can flow over long distances (>300 km).

3.5 Structures and Textures of Magmatic Rocks

The description of magmatic rocks involves examining various aspects, including the crystal size and shape and the geometric arrangement of the individual crystals as well as noting the presence or absence of a number of other features.

Table 3.2 Crystal sizes in magmatic rocks. (After Jerram and Petford 2011)

Crystal size	Features
Fine grained (aphanitic/hyaline for glassy rocks; <1 mm)	Rare crystal boundaries observed in the field or distinguishable with a hand lens
Medium grained (phaneritic; 1–5 mm)	Most crystal boundaries can be distinguished with a hand lens
Coarse grained (phaneritic; >5 mm)	Almost all crystal boundaries can be distinguished with the naked eye

Crystal Size

Crystal size is mainly influenced by the cooling rate of the magma. Coarse crystalline rocks are called **phaneritic**, while fine crystalline rocks are termed **aphanitic**. The former are generally intrusive, while the latter are mostly extrusive (Table 3.2). A mixed form, in which larger crystals (**phenocrysts**) occur in a finer-grained matrix (**groundmass**), is termed a **porphyritic structure.** A **porphyry** is a rock that contains 50% of large, well-developed **phenocrysts** (often plagioclase or alkali feldspar), which are distributed in a fine-grained matrix. A coarse-grained igneous rock is termed a **pegmatite** (pegmatitic texture; usually—but not always—granitic, often with crystals > c. 1.0 cm in diameter) with the fine-grained equivalent being an **aplite** (aplitic texture; granitic composition with crystals <1 mm in diameter). Even finer is the glassy volcanic rock **obsidian (pitchstone)**, which is formed by rapid cooling of a magma.

Magmatic rocks may also contain foreign inclusions consisting of rocks (**xenoliths**) or minerals (**xenocrysts**) that were derived from the surrounding rock during the ascent of the magma.

Crystal Shape

Mineral shapes within a magmatic rock can be described as follows (Fig. 3.4):

- **Idiomorphic**—crystals with a well developed crystal shape, with distinct crystal faces and a characteristic geometry.
- **Hypidiomorphic**—crystals that resemble their ideal form, but whose growth was restricted by other minerals
- **Xenomorphic**—crystals that are very irregular in shape and do not give any indication of their ideal crystal form.

In the coarser magmatic rocks the majority of crystals are idiomorphic to hypidiomorphic in form. In volcanic rocks, **phenocrysts** and some accessory minerals (e.g., zircon, apatite) can be idiomorphic because they crystallized early.

Texture

Texture refers to the geometrical arrangement and distribution of mineral grains within a rock and is strongly influenced by the order in which they crystallize. A **trachytic texture** is a flow texture which describes the parallel/subparallel ar-

3

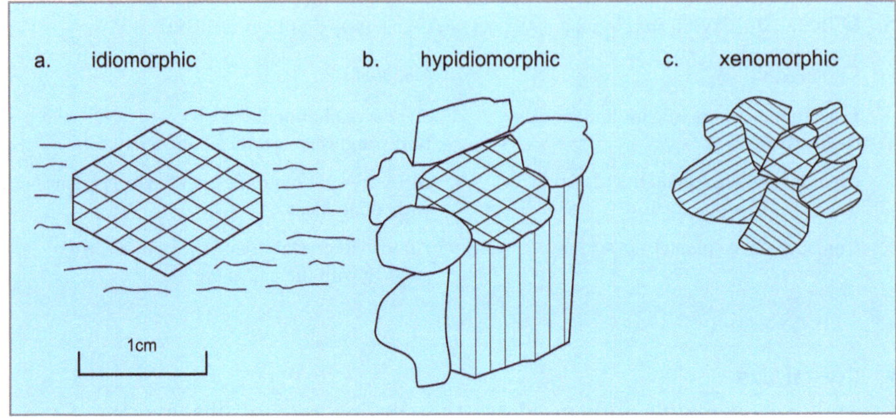

Fig. 3.4 Crystal shapes in magmatic rocks. (After Jerram and Petford 2011)

rangement of elongated plagioclase crystals (phenocrysts) within the fine-grained groundmass.

Vesicles (bubbles) are gas bubbles trapped in solidifying magma. **Vesicular texture** is usually formed in volcanic rocks (e.g., vesicular basalt, pumice). If the cavities (amygdules/amygdales) are subsequently filled with secondary minerals, this is referred to as an **amygdaloidal texture** (e.g., amygdaloidal basalt).

3.6 Classification of Magmatic Rocks

The proportion of minerals in a magmatic rock is an important feature that is used for comparison and classification purposes (Table 3.3). Magmatic rocks can generally be divided into three to four groups based on their chemistry and color (dependent on the mineral content):

- **Acidic/Felsic rocks**—light-colored rocks containing mainly quartz, feldspar, and feldspathoid. The SiO_2 content is >65% silica.
- **Basic/Mafic rocks**—mostly dark colored and contain iron- and magnesium-rich minerals such as pyroxenes, amphiboles and olivines. The SiO_2

Table 3.3 Classification of magmatic rocks based on their SiO_2 contents. (After Thorpe and Brown 1985)

Geochemical term	Definition weight % SiO_2	Color
Acidic	>65	Light
Intermediate	52–65	Medium
Basic	45–55	Dark
Ultrabasic	<45	Dark

content is ≤55% silica. Rocks that contain only iron- and magnesium-rich minerals are called ultrabasic/ultramafic (<45% silica).

- **Intermediate rocks**—are located between the two end members above in terms of their mineral content and contain both light- and dark-colored minerals; they are intermediate in color.

The color of the rock and the minerals present can be combined with the grain size to provide a simple field classification for magmatic rocks (◘ Fig. 3.5).

The **IUGS classification system** is a more detailed scheme to classify magmatic rocks. Classification is in two stages (◘ Figs. 3.6 and 3.7):

1. What is the grain size of the rock, that is, whether the rock is phaneritic (i.e. coarse-grained—plutonic) or aphanitic (i.e. fine-grained—volcanic)?
2. Determine the proportions of the five most common minerals or mineral groups in the rock. When looking at hand samples, these minerals are best identified using the following properties (see also ▶ Chap. 2 for more detailed descriptions):

- **Quartz**—translucent, vitreous luster, absence of obvious cleavage, conchoidal fracture.
- **Plagioclase**—cleavage, twinning on the cleavage faces.
- **Alkali feldspar**—cleavage, Carlsbad twins are often visible in hand specimens, often pink to brownish color.
- **Mafic minerals**—black, brown, or green colors; 90° cleavage (pyroxenes), 120° cleavage (amphiboles).

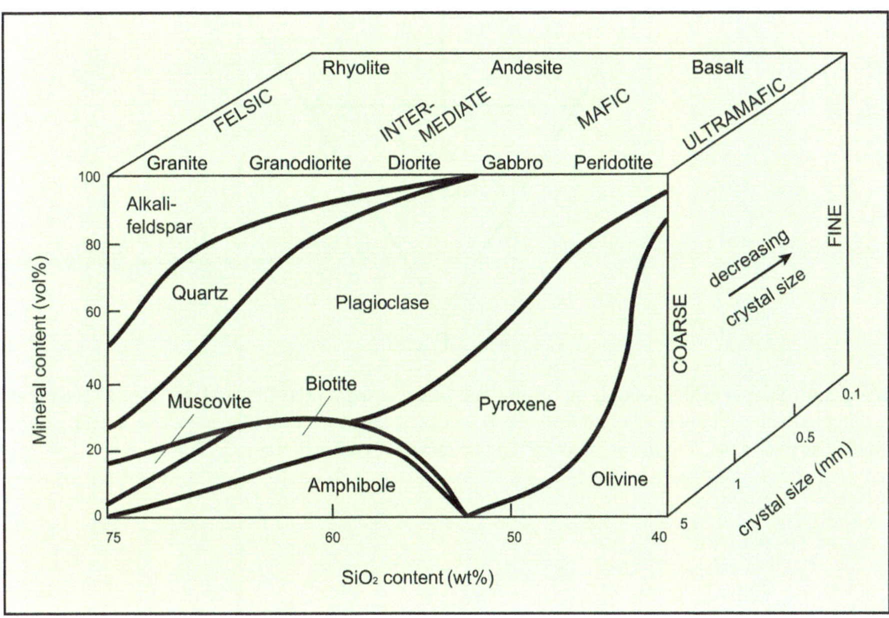

◘ **Fig. 3.5** Simple field classification, based on color (light vs. dark minerals), important silicate minerals and grain size. (After Jerram and Petford 2011)

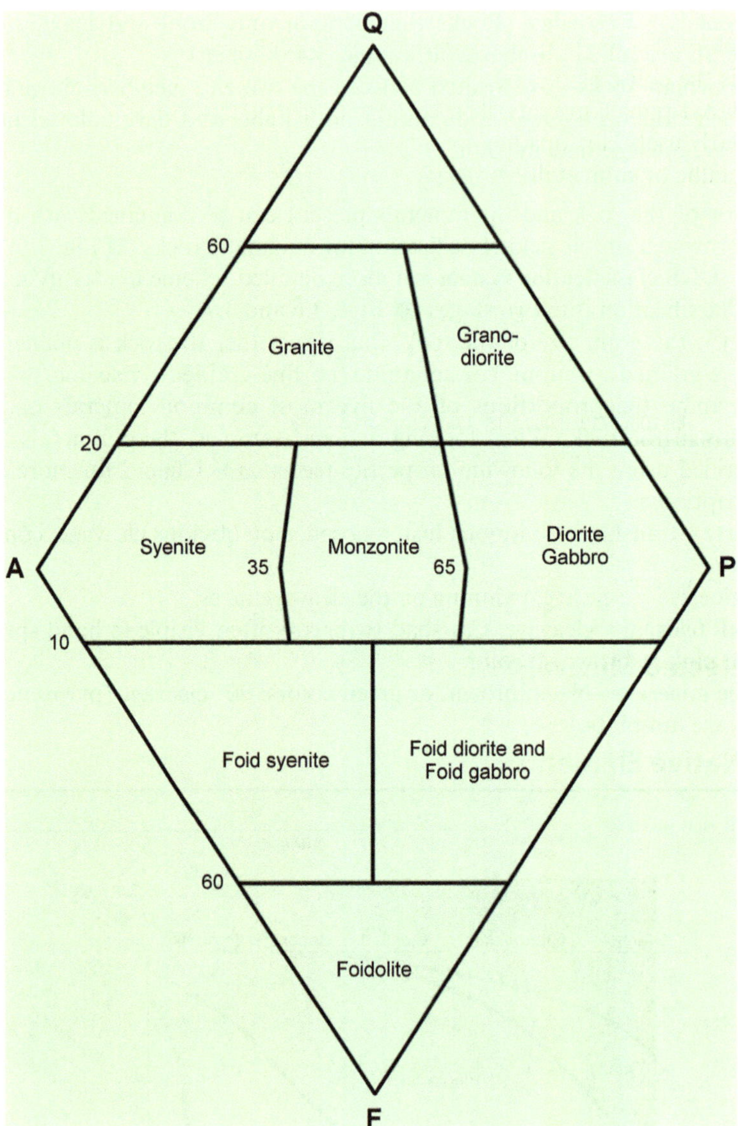

Fig. 3.6 Simple field classification of plutonic rocks, based on the IUGS-classification scheme, according to their mineralogical compositions: *Q* quartz; *A* alkali feldspar; *P* plagioclase; *F* feldspathoid (foid). The rock must contain less than 90% mafic minerals. (After Blatt et al. 2006)

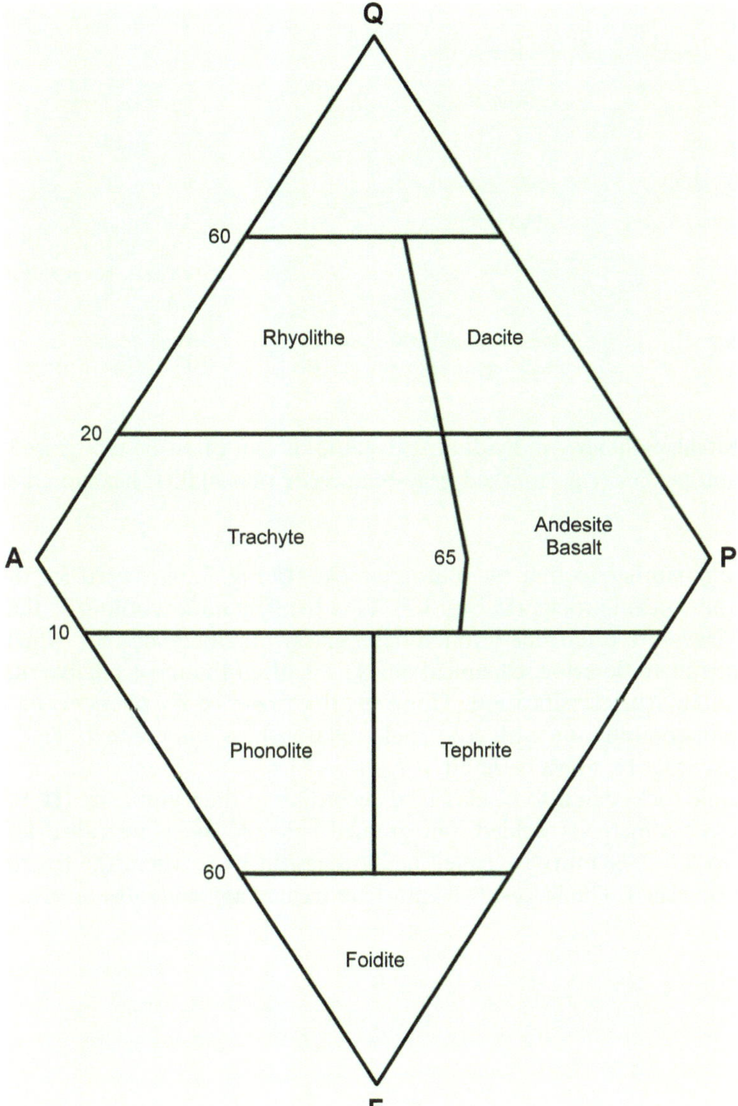

Fig. 3.7 Simple field classification of volcanic rocks, based on the IUGS-classification, of volcanic rocks according to their mineralogical compositions: *Q* quartz; *A* alkali feldspar; *P* plagioclase; *F* feldspathoid (foid). The rock must contain less than 90% mafic minerals. (After Blatt et al. 2006)

3

Table 3.4 Grain size based nomenclature of the most common volcanoclastic rocks. (After McCann and Valdivia Manchego 2015)

Grain size	Non-consolidated tephra	Pyroclastic rock
<0.063 mm	Fine-grained ash	Fine tuff
0.063–2 mm	Coarse-grained ash	Coarse tuff
2–64 mm	Lapilli tephra	Lapillistone (or lapilli tuff/breccia) or volcanic conglomerate/breccia
>64 mm	Bomb tephra (molten ejecta) Block tephra (solid/angular ejecta)	Agglomerate Pyroclastic breccia

- **Foids/feldspathoids**—individual feldspathoids can often be recognised by specific properties (e.g., marked gray-blue color of sodalite, hexagonal shape of leucite).

A diagram, similar to that for plutonic rocks (Fig. 3.6), is used for the classification of volcanic rocks (Fig. 3.7). Due to their rapid cooling at the Earth's surface, lavas are often fine-grained (e.g., glassy, microcrystalline), which makes them difficult to describe. Chemical analysis is often the most reliable method to provide a correct classification. However, the presence of phenocrysts (usually 10–50%) in combination with color, field relationships and contexts, etc., can also be used to describe a lava in detail.

Volcanic rocks can also be classified according to their grain size (Table 3.4). Compacted, sometimes welded, fine-grained volcanic rocks are called **tuffs** while coarser rocks (2–64 mm) are called **lapillistones**/tuffs (or where the fragments are angular, breccias). The largest (>64 mm) are termed **agglomerates** or **volcanic breccias**.

3.7 Plutonic Rocks

▪ Granite

a Granite (scale 2 cm) **b** Rapakivi granite

Color – mottled appearance, white, gray, pink, and red

Grain size – coarse to very coarse

Texture – granular rock, very homogeneous (sometimes banded), often porphyritic (phenocrysts mostly feldspar); xenoliths often present; dikes and veins of micrograinte may occur

Mineralogy – light-colored minerals (80–100%), alkali feldspar (35–100%), plagioclase (0–65%), and quartz (min. 10%, normally 20–60%). Accessory minerals include biotite (also muscovite), augite, hornblende, tourmaline, topaz, apatite, titanite (sphene), zircon, ilmenite, pyrite, and magnetite

Occurrence – found in large intrusions (e.g., batholiths, stocks), but also sills and dikes. Volcanic counterpart is rhyolite

Varieties

- **Rapakivi granite**—a porphyric hornblende granite with round to ovoid alkali feldspar crystals (orthoclase; 2–3 cm diameter) surrounded by plagioclase
- **Alkali feldspar granite**—light-colored granite comprising mainly alkali feldspar (<10% plagioclase). Accessory minerals include augite, hornblende, and zircon
- **Augite-Hornblende granite**—dark-colored granite due to presence of augite and hornblende
- **Biotite granite**—large amount of biotite (up to 20%); a **two-mica granite** would contain a significant proportion of muscovite
- **Tourmaline granite**—large amount of tourmaline
- **Aplite**—very fine-grained granite (often occurs as dikes), rare mafic minerals

▪ Granodiorite

Granodiorite (scale 2 cm)

Color – predominantly gray (the higher the proportion of mafic components, the darker the rock).

Grain size – as for granite

Texture – granular, similar to granite

Mineralogy – difficult to distinguish from granite, but usually looks darker. Mineralogy is similar to granite, although with different feldspar ratios (plagioclase 65–100%, alkali feldspar 0–35%); it may be transitional with granite

Occurrence – the most abundant of the granitic rocks, often intrusive (e.g. batholiths), and found within granitic massifs. The volcanic equivalent is dacite

Varieties

- **Trondhjemite**—quartz-rich variety (>20%) with little or no alkali feldspar; dark minerals (<15%) mostly biotite and hornblende.
- **Tonalite**—feldspar mostly plagioclase and usually no alkali feldspar; quartz about 20%; dark minerals (10–40%) mostly biotite and hornblende (both often porphyric).

Syenite

Hornblende syenite (scale 2 cm)

Colour – light to dark grey, also red or white

Grain size – medium to coarse (sometimes pegmatitic)

Texture – similar to granite

Mineralogy – rich in feldspars and low in quartz compared to granites. Light-colored minerals (60–100%), of which 80–100% are feldspar (alkali feldspar 65–100% and plagioclase 0–35%), quartz (0–20%) or feldspathoid (0–10%), these last two being mutually exclusive; accessory minerals include biotite, pyroxene, fluorite, zircon, titanite, apatite, ilmenite, and magnetite

Occurrence – not particularly common, mostly associated with granites or present in small intrusions

Varieties

- **Alkali syenite**—almost plagioclase-free; often found associated with alkali granites

■ Monzonite

3

Monzonite (scale 2 cm)

Color – light to dark gray, also greenish, brownish, and red

Grain size – generally medium

Texture – granular, occasional flow structures (oriented minerals), and/or tabular alkali feldspar crystals

Mineralogy – rich in feldspar and low in quartz (same as syenite). Plagioclase > alkali feldspar. Light-colored minerals (55–90%), of which 80–100% are feldspar (alkali feldspar 35–65% and plagioclase 35–65%), rare quartz (0–20%), and feldspathoid (0–10%); accessory minerals include pyroxene, hornblende, biotite. In transitional (to diorite/gabbro) forms, plagioclase predominates, quartz is < 5% and pyroxene up to 20%.

Occurrence – associated with granites and granodiorites

▪ Foid-Syenite

Nepheline syenite (scale 2 cm)

Color – light colored (gray, pink), also dark green

Grain size – medium to coarse; sometimes with feldspar phenocrysts

Texture – granular, sometimes oriented crystals, including tabular alkali feldspar and elongate hornblende

Mineralogy – no quartz, mainly feldspathoid (especially nepheline) and alkali feldspar. Light-colored minerals 55–100%, including 40–90% feldspar (alkali feldspar 50–100%, plagioclase 0–50%), and feldspathoid (10–60%, typically: nepheline, sodalite, leucite); accessory minerals include biotite, pyroxene, and amphibole

Occurrence – often associated with syenites; occurs as small intrusive bodies (e.g., stocks, sills)

3

■ Diorite

Diorite (scale 2 cm)

Color – black and white speckled appearance, sometimes dark green or pink

Grain size – coarse, but very variable (sometimes pegmatitic), sometimes phenocrysts (e.g., hornblende)

Texture – granular, often containing xenoliths, sometimes foliation

Mineralogy – mainly plagioclase and mafic minerals (e.g. amphibole, pyroxene). Light-colored minerals 50–85%, of which 80–100% are feldspar (plagioclase—mostly oligoclase, andesine 65–100%, alkali feldspar 0–35%), quartz 0–20%, or feldspathoid 0–10%. Dark-colored minerals 15–50%, including biotite and/or pyroxene. Accessory minerals include apatite, titanite (sphene), iron oxides, zirconium, and garnet. With 5–20% quartz = quartz diorite, >20% quartz = tonalite.

Occurrence – small intrusive bodies, grading laterally to granites, or gabbros. Volcanic counterpart is andesite

Gabbro

Gabbro (scale 2 cm)

Color – gray, dark gray, black, greenish, bluish

Grain size – coarse, sometimes pegmatitic

Texture – granular, often layered (light and dark minerals), with individual layers ranging in thickness from centimeters to several meters

Mineralogy – mainly mafic minerals (e.g. pyroxene, olivine) and plagioclase. Looks darker than diorite. Light-colored minerals 55–80%, of which 80–100% are feldspar (plagioclase: darker varieties, labradorite, bytownite 65–100%, alkali feldspar 0–35%), quartz (1–20%), feldspathoid (0–10%). Dark-colored minerals 20–65% especially pyroxene (augite), hornblende, olivine and/or biotite. Accessory minerals include apatite, pyrite, magnetite, ilmenite, and serpentine

Occurrence – an intrusive rock found in stocks, and sometimes lopoliths. Often associated with pyroxenites and anorthosites. Volcanic counterpart is basalt

Varieties

- **Norite**—dark gray with hypersthene. Contains orthopyroxene or pigeonite instead of augite
- **Troctolite**—contains olivine instead of augite
- **Essexite**—fine- to medium-grained, sometimes porphyritic. High proportion of pyroxene (and, therefore, almost black)

Anorthosite

Color – gray to white

Grain size – medium to coarse

Texture – granular, sometimes oriented crystals, sometimes layered

Mineralogy – mainly plagioclase (>90% oligoclase/andesine to bytownite). Accessory minerals include pyroxene, olivine, and iron oxides

Occurrence – found in both large (e.g., stocks, batholiths) and smaller intrusive bodies, often layered and associated with gabbros. Also as bodies (over hundreds of km^2) within metamorphic areas

▪ Pyroxenite

Color – green, dark green to black

Grain size – medium to coarse

Texture – granular, sometimes layered

Mineralogy – ultramafic rock consisting mainly of pyroxene (clinopyroxene or orthopyroxene), olivine, hornblende, iron oxides, or biotite. Feldspars are rare/absent (unlike gabbro) and the percentage of olivine is <40% (unlike peridotite).

Occurrence – intrusive bodies (stocks, dikes) or as bands within stratified gabbros

▪ Kimberlite

Color – bluish, greenish, or black

Grain size – amorphous or fine-grained, sometimes phenocrysts

Texture – often porphyritic, xenoliths are common

Mineralogy – ultramafic volcanic rock comprising mainly olivine (sometimes serpentinized), mica (phlogopite) with garnet (pyrope) and orthopyroxene. Accessory minerals include ilmenite, spinel, rutile, calcite, chromite, and diamond

Occurrence – found mainly in kimberlite pipes (hundreds of meters in diameter) which are the main source of diamonds; less commonly as dikes

3.8 Volcanic/Subvolcanic Rocks

Rhyolite

Rhyolite (scale 2 cm)

Color – white, gray, greenish, reddish, or brownish. Sometimes banded

Grain size – glassy to fine

Texture – often layered or with flow structures, showing variations in grain size or color. Oriented phenocrysts (quartz, feldspar, hornblende, mica) are sometimes present. Vesicular or amygdaloidal textures are sometimes present. Occasional spherulites (radial growth of quartz/feldspar needles).

Mineralogy – light-colored minerals (80–100%), of which 20–60% quartz, 40–80% feldspar (alkali feldspar 35–100%, plagioclase 0–65%). Dark-colored minerals 0–20%—pyroxene, biotite, zircon, and apatite

Occurrence – found in volcanic areas (volcanic vent plugs, lava flows, dikes)

Varieties

- **Quartz porphyry**— quartz phenocrysts and occasionally biotite present
- **Granite porphyry**—phenocrysts of alkali feldspar, quartz and sometimes biotite and/or plagioclase present

Microsyenite

Color – gray, reddish, brownish

Grain size – medium

Texture – granular, often with phenocrysts (mostly alkali feldspar)

Mineralogy – similar to syenite (alkali feldspar, with biotite, hornblende, pyroxene, or quartz)

Occurrence – found in dikes and sills associated with intrusions of syenite or nepheline syenite; also associated with trachyte

▪ Trachyte

3

Trachyte (scale 2.2 cm)

Color – gray; also brownish, pink, or yellowish

Grain size – fine

Texture – often porphyritic (phenocrysts of sanidine, but also plagioclase, hornblende, pyroxene), with flow structures (trachytic texture)

Mineralogy – feldspar rich with alkali feldspar > plagioclase. Light minerals 60–100%, of which 80–100% are feldspar (alkali feldspar 65–100%, plagioclase 0–35%), quartz (0–20%) or feldspathoid (0–10%). Dark minerals 0–40% (including pyroxene, hornblende, biotite)

Occurrence – occur as lava flows, often associated with basalt and small intrusive bodies (dikes, sills)

▪ Microdiorite

Color – gray to dark gray, sometimes greenish or pink

Grain size – medium

Texture – usually porphyritic (phenocrysts of hornblende, biotite or augite)

Mineralogy – as for diorite

Occurrence – found in intrusive bodies (dikes, sills), often found in swarms around diorite or granite intrusions

▪ Andesite

Trachyandesite

Color – gray, brown, greenish or almost black

Grain size – fine, glassy/amorphous

Texture – flow structure, usually porphyritic (phenocrysts of plagioclase, pyroxene, hornblende or biotite), vesicular or amygdaloidal

Mineralogy – feldspar rich (plagioclase > alkali feldspar), quartz, pyroxene, amphibole and biotite. Light-colored minerals 60–85%, of which 80–100% are feldspar (plagioclase 65–100%, alkali feldspar 0–35%), quartz (0–20%) or feldspathoid (0–10%). Dark-colored minerals 10–40% (including biotite, augite, hornblende, olivine, magnetite, zircon). Plagioclase in the fine-grained groundmass is usually oligoclase-andesine

Occurrence – found in lava flows, stocks and dikes. Often associated with basalts, dacites, trachytes and rhyolites

Basalt

3

Basalt (scale 2 cm)

Color – black, dark gray, may weather to reddish or greenish

Grain size – fine, sometimes glassy/amorphous

Texture – sometimes porphyritic (phenocrysts of hornblende, pyroxene, olivine, plagioclase), vesicular or amygdaloidal (amygdules filled with zeolites, carbonates or quartz). Xenoliths are sometimes present (often olivine or pyroxene). Columnar jointing may be present

Mineralogy – mainly plagioclase and pyroxene. Dark-colored minerals 40–70% (pyroxene, olivine, magnetite, ilmenite, biotite), light-colored minerals 30–60%, of which 80–100% are feldspar (plagioclase 65–100%, alkali feldspar 0–35%), quartz (0–20%), or feldspathoid (0–10%).

Occurrence – commonly found as lavas, flood basalts, dikes, and sills

Varieties

- **Diabase/Dolerite**—coarse, often with phenocrysts of plagioclase
- **Tholeiitic basalt**—calcic plagioclase, olivine is rare/absent
- **Alkali basalt/Alkali olivine basalt**—with olivine, augite and nepheline
- **Olivine basalt**—with olivine and pyroxene phenocrysts

Dacite

Dacite (scale 2.2 cm)

Color – gray, brownish, yellowish

Grain size – fine, amorphous

Texture – often porphyritic (phenocrysts of zoned plagioclase, quartz, hornblende, biotite, rarely alkali feldspar), sometimes flow structures

Mineralogy – quartz- and plagioclase-rich rock. Light-colored minerals 70–95%, of which 20–60% quartz, 40–80% feldspar (plagioclase 65–100%, K-feldspar 0–35%). Dark-colored minerals (5–30%)—pyroxene, hornblende, biotite, zircon, and magnetite

Occurrence – found in dikes, sills, and lava domes

Obsidian

Obsidian (scale 1.9 cm)

Color – black, dark brown, gray.

Grain size – amorphous

Texture – amorphous, rare phenocrysts (more often in pitchstone, quartz, feldspar). Snowflake obsidian is characterised by the presence of spherulites (radially clustered crystals of cristobalite which form as a result of devitrification). Vitreous. Conchoidal fracture

Mineralogy – variable, but mostly like rhyolite. However, there are also trachytic, andesitic, and phonolithic obsidians

Occurrence – mostly in the marginal areas of rhyolithic lava flows. Pitchstone is similar, but has a higher water content, a hackly fracture and a resinous luster

▪ Pumice

Pumice (scale 2 cm)

Color – white, gray, yellowish, may be darker

Grain size – fine/amorphous

Texture – highly vesicular (up to 85% of rock), vesicles are irregular/oval

Mineralogy – composition is similar to rhyolite, but can also be dacitic, andesitic, trachytic, or phonolithic. Basaltic pumice is also known

Occurrence – formed in explosive eruptions, in viscous lavas by exsolution (degassing) of volcanic gases, due to pressure release

Scoria

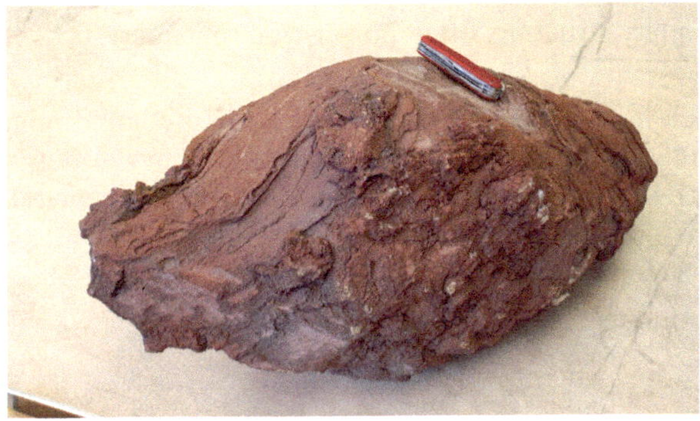

Scoria, spindle-shaped bomb (scale 9 cm)

Color – dark brown, black, reddish

Grain size – fine/amorphous

Texture – denser than pumice, with larger vesicles and thicker vesicle walls; sometimes with phenocrysts

Mineralogy – variable, same as basalt or andesite

Occurrence – forms due to exsolution (degassing) of volcanic gases (lavas are less viscous than those which form pumice). Can build small cones (scoria cones) or may be ejected as bombs/blocks or lapilli.

Phonolite

Phonolite (scale 2 cm)

Color – dark green to gray

Grain size – fine to mixed fine-coarse

Texture – often platy structure, often porphyritic (phenocrysts of feldspar, nepheline)

Mineralogy – rich in feldspar and feldspathoid. Light-colored minerals 60–100%, of which 40–90% are feldspar (alkali feldspar 50–100%—often sanidine, plagioclase 0–50%), feldspathoid (10–60%, nepheline, sodalite, leucite). Dark-colored minerals 0–40%—biotite, pyroxene, amphibole, olivine, ilmenite, apatite, titanite, magnetite, and zircon

Occurrence – found as lava flows, sills, dikes. Often associated with trachyte and nepheline syenite

3.9 Pyroclastic Rocks

▪ Agglomerate

Agglomerate (top) and fault (scale 10 cm)

Color – dark to light (depending on the composition)

Grain size – angular to rounded fragments (>64 mm diameter) in a fine-grained matrix

Texture – fragments are variable in size—from blocks to bombs

Mineralogy – very variable and dependent on the source rock (e.g., basalt, andesite)

Occurrence – found in the proximal area of volcanoes (craters, rims), associated with tuffs and lava flows. Volcanic breccia—angular fragments/clasts; agglomerate—rounded fragments/clasts

■ Ashes and tuff

Tuff (scale 2 cm)

Color – dark to light (depending on composition)

Grain size – fine (<2 mm diameter)

Texture – tuff is consolidated ash, often laminated/bedded like sediment, may show grading. Often includes lithic fragments (e.g., rhyolites, andesite = lithic ash/tuff), glassy fragments (e.g., pumice = vitreous ash/tuff), or crystals (e.g., feldspar, hornblende = crystal ash/tuff). Sometimes with lapilli (round to elliptical fragments, 2–64 mm diameter = lapilli tuff). Fragments (e.g., pumice) may be compressed (fiamme) and are typical of welded tuffs.

Mineralogy – very variable and dependent on the source rock (e.g., basalt tuff, rhyolite tuff, andesite tuff, trachyte tuff)

Occurrence – airborne ash, transported by eruption column and wind. Coarse-grained material is deposited proximally, but fine-grained dust can be transported hundreds of kilometers. Associated with lava flows, agglomerates, and sediments. Ash and tuff layers can form important marker horizons

▪ Ignimbrite

Non-welded ignimbrite

Color – variable, gray, reddish, yellowish, brown, black (depending on composition and density)

Grain size – amorphous/glassy, rare phenocrysts (biotite, quartz, sanidine, hornblende, rarely pyroxene)

Texture – matrix of glass shards/ash with fragments of pumice (lapilli, <1 cm diameter), glass or crystals. Fragments of pumice can be flattened (fiamme), especially in the lower part of the flow. Often layered. Columnar jointing sometimes present.

Mineralogy – variable—depending on initial magma composition (e.g. dacite, rhyolite, rarely basalts)

Occurrence – deposited from highly concentrated pyroclastic flows (glowing clouds—*nueés ardentes*)—mixtures of gas and fragments (ash and pumice lapilli). Associated with lava flows, agglomerates, and sediments

3.10 Ultramafic Rocks

■ Dunite

Dunite (scale 2 cm)

Color – green, brownish

Grain size – coarse

Texture – granular, sometimes layered

Mineralogy – olivine (>90%), pyroxene, chromite, magnetite, and garnet.

Occurrence – found as cumulates of olivine and pyroxene in basaltic magma chambers. Present in layered gabbros or anorthosites associated with peridotite

■ Peridotite

Peridotite (scale 2 cm)

Color – greenish to black

Grain size – medium to coarse

Texture – coarse, sometimes layered, rarely porphyritic

Mineralogy – olivine (40–90%), pyroxene and/or amphibole. Accessory minerals include biotite, chromite, garnet, and plagioclase. Harzburgite is a pure olivine-orthopyroxene rock, without augite

Occurrence – dominant rock in the Earth's mantle. Probably forms as a cumulate of olivine crystals in a basaltic magma. Found in intrusive bodies (dikes, small stocks) and forming oceanic crust (lowermost part of an ophiolite sequence), also as part of stratified gabbro intrusions (with pyroxenite and anorthosite). Occurs as xenoliths in basalts

Metamorphic Rocks

Contents

T. McCann, *Pocket Guide Geology in the Field*,
https://doi.org/10.1007/978-3-662-63082-2_4

Metamorphosis, in the geological sense, comprises the various mineralogical, structural, and chemical transformations that occurs when rocks within the Earth's crust are subjected to elevated temperatures (between diagenesis at max. c. 200 °C and melting at c. 850 °C) or pressures, or contact with hot mineral-rich fluids (or more usually, a combination of these factors). Metamorphism resulting from increasing temperatures and pressures is termed **prograde metamorphism**, with the opposite being termed **retrograde metamorphism**. The original rock (whether magmatic, sedimentary, or metamorphic in origin) is called a **protolith**.

Metamorphism occurs at different scales and is generally classified as regional or local metamorphism.

Regional metamorphism:

- Orogenic metamorphism—occurs in areas of mountain building (orogeny) during crustal deformation associated with plate convergence.
- Burial metamorphism—occurs in rocks which have been buried under significant thicknesses of sediments or volcanic rocks.
- Ocean-ridge metamorphism—is associated with the presence of mid-ocean ridges.

Characteristic series of metamorphic rocks exist in areas that have undergone regional metamorphism. They typically cover large areas (10^2–10^3 km^2).

Local metamorphism:

- Hydrothermal metamorphism—occurs in zones where hydrothermal and other fluids circulate (**metasomatism**).
- Impact metamorphism—occurs due to the impact of an extraterrestrial body (**shock metamorphism**)
- Contact metamorphism—occurs in the host/country rock surrounding magmatic intrusions (for granitic intrusions at temperatures of around 800–850 °C, for gabbro or diorite intrusions the temperatures are 900–1100 °C). Extensive contact aureoles (up to approx. 10 km) can form around large intrusions.

The most common form of local metamorphism occurs in contact zones of magmatic intrusions. Here thermal energy is transferred directly from the intrusive rocks to the host rock. The presence of magmatic fluids can also result in extensive metasomatism around granitic intrusions. In large aureoles, mineralization zones with clear proximal (high-grade minerals) to distal (low-grade minerals) relationships can form. Contact metamorphism may also be restricted to small areas (centimeters to meters), such as around sills or dikes.

4.1 Metamorphic Facies

The metamorphic facies scheme helps to describe and classify metamorphic rocks on the basis of their mineralogy (◘ Fig. 4.1). In areas which have undergone metamorphism, minerals will recrystallize to new minerals as a result of changing pressure and temperature conditions. Indeed, there is a direct relationship

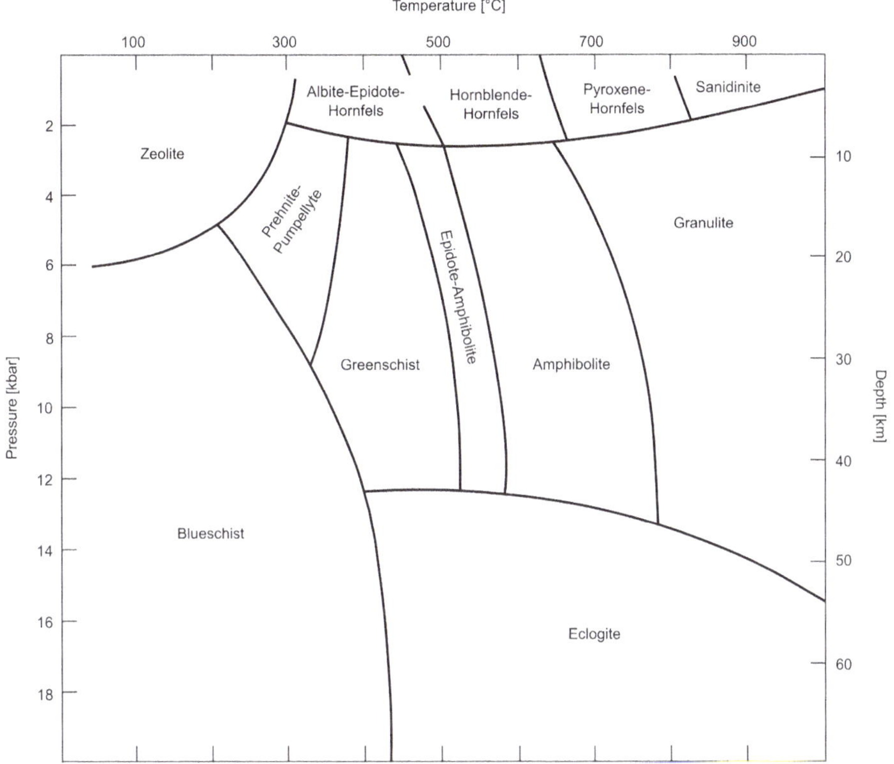

Fig. 4.1 Distribution of metamorphic facies as a function of pressure (depth) and temperature (after Blatt et al. 2006)

between the minerals formed and the metamorphic conditions (e.g. temperature, pressure, fluids). Thus, each metamorphic facies can be defined by a suite of characteristic minerals.

Twelve facies are currently recognised ranging from the **zeolite facies** (low pressure, low temperature) to the **eclogite facies** (high pressure). (Fig. 4.1). However, for field purposes it is easier to focus on recognising the key minerals and establishing the main pressure/temperature conditions under which metamorphism took place (Fig. 4.1, Table 4.1).

4.2 Structures and Microstructures of Metamorphic Rocks

Variations in the structure and composition of metamorphic rocks depend both on the changing pressure and temperature conditions and on the original parent rock. The role of fluids (i.e., metasomatism) is also important. Processes such as pressure solution, melting, and tectonic deformation form the starting point for a number of characteristic structures. These are often linear/planar in form,

Table 4.1 Common metamorphic rock compositions and the key minerals present within them. The minerals present correspond to particular metamorphic conditions (i.e. pressure/temperature) (after Argles in Coe 2010)

Pressure/temperature conditions	Ultramafic rocks (peridotite)	Mafic rocks (basalt)	Felsic rocks (granite)	Pelitic rocks (mudstone)	Calcareous rocks (impure limestone)
Low temperature	Serpentine	Chlorite	Chlorite, epidote	Chlorite	Talc
Low pressure		Pyroxenes, olivine No garnet	Andalusite	Andalusite, cordierite No garnet	
Medium pressure/temperature	Talc (common)	Actinolite, epidote, zoisite		Chloritoid, staurolite	Tremolite
High pressure		Lawsonite, Na-pyroxene, rutile, glaucophane No plagioclase	Na-pyroxene, kyanite No plagioclase		Zoisite
High temperature	Orthopyroxene	Clinopyroxene, orthopyroxene	Orthopyroxene, cordierite, sillimanite	Sillimanite, spinel, orthopyroxene No muscovite	Wollastonite, Mg-olivine, Ca-pyroxene, spinel
Wide pressure and temperature ranges	Olivine, chlorite, magnesite	Garnet, hornblende, plagioclase, biotite, quartz, titanite	Quartz, biotite, K-feldspar, plagioclase, muscovite	Muscovite, biotite, garnet, quartz, plagioclase	Calcite, dolomite, plagioclase, Ca-garnet, hornblende, chlorite, epidote

and thus, metamorphic rocks can be broadly subdivided into **foliated** (e.g., schist, gneiss) and **non-foliated** rocks (e.g., quartzite, marble).

▪ Lineation

Lineation forms in areas of ductile deformation, where minerals and mineral aggregates (e.g., hornblende, tourmaline) within the rock become oriented in the principal deformation direction.

Mineral lineation, Monte Rosa Nappe, Italian Alps (scale 2 cm; Photo N. Froitzheim)

Foliation

Foliation is a general term referring to laminar/planar layering in metamorphic rocks. These planar structures form as a result of deformation and result from the parallel alignment of sheet silicate minerals (e.g., schist) and/or compositional layering (e.g., gneiss) and may represent original features that have been overprinted by deformation

Cleavage is a parallel foliation (layering) typical of fine-grained or low-grade metamorphic rocks where minerals show a preferred orientation (e.g., slatey cleavage). Cleavage planes are generally not parallel to bedding, but formed as a result of deformation.

Cleavage, Val Trupchun, Switzerland. (Picture width 20 cm; Photo N. Froitzheim)

Schistosity is a type of foliation characteristic of coarser and intermediate- to high-grade metamorphic rocks in which platy minerals (e.g., mica, chlorite) show a preferred orientation (e.g., mica schist).

Schistosity

Gneissic texture occurs when the foliation of a rock consists of millimeter to centimeter thick bands in which the mineral ratios, colors, or textures vary. Gneiss is characterized by its banding, a foliation that develops during dynamic metamorphism which results in the formation of dark- and light-colored mineral bands. Where partial melting occurs, a **migmatite** (transitional between a metamorphic and a magmatic rock) may form.

Kink bands form due to the deformation/folding of foliation planes. They form in clay-rich shales as a result of lateral compression which leads to buckling.

Kink bands, Alpbach, Austria (scale 30 cm; Photo K. Wellnitz)

Shear zones form as a result of localized deformation (ductile, brittle, or mixed) which is concentrated in a planar zone between rocks which are less deformed.

Shear zone in sandstone

4.3 Describing Metamorphic Rocks

The high temperatures and pressures that occur in rocks during metamorphosis result in the formation of a number of characteristic features.

Mineral Growth

One major effect of the elevated temperatures is the increase in grain size in most rocks as a result of recrystallization.

Growth of Porphyroblasts

Large, generally euhedral, crystals which develop as a result of metamorphism are termed porphyroblasts. These are broadly the metamorphic equivalents of phenocrysts in magmatic rocks.. The particular arrangement of the porphyroblasts in the rock forms a characteristic structure. The following terms are important:

- **Porphyroblastic texture**—a granular texture where the large crystals (porphyroblasts) are significantly larger than the fine-grained groundmass.
- **Granoblastic texture**—a texture where the individual crystals are approximately equigranular and there are no porphyroblasts.
- **Augen**—are eye-shaped porphyroblasts or porphyroclasts (fragments of the original rock, prior to metamorphism) of minerals or mineral aggregates in rocks which have undergone metamorphism and shearing. They are surrounded by microshears and may rotate during deformation.

Augengneiss, Austria (scale 16 cm; Photo N. Froitzheim)

▪ Boudinage

Sausage-shaped structures (boudins) formed in porphyroblasts/clasts due to shearing and stretching.

Boudinage, Pohorje, Slovenia (scale 2.5 cm; Photo N. Froitzheim)

▪ Mylonitization (reduction in grain size)

In areas of pervasive ductile deformation, grain size may be reduced during metamorphism. This process is termed mylonitization, and the resultant rocks are called mylonites.

Mylonite, Moine Thrust, Scotland (scale 30 cm; Photo N. Froitzheim)

4.4 Recognition and Classification of Metamorphic Rocks

The classification of metamorphic rocks is based on the source rock (protolith) and the degree of deformation. Metamorphic rocks (e.g., slate, rock, gneiss) are given an appropriate prefix if they have a characteristic composition, for example, garnet gneiss, mica schist. In addition, there are historical names for certain metamorphic rocks (e.g., marble).

▪ Protolith

There are three main groups of metamorphic rocks based on the parent rock; ranging from clastic sedimentary rocks (e.g., sandstones), calcareous rocks (e.g., limestones), and mafic or intermediate volcanic or pyroclastic rocks (e.g., basalts, andesites). Other metamorphic rocks, which do not belong to the these groups, include granulite, serpentinite, skarn, and mylonite (◘ Table 4.2).

▪ Texture

The microstructure (including crystal/grain size) often provides an important indication of the intensity (degree) and type of metamorphism (◘ Table 4.3). Very fine-grained rocks are usually found in areas of low-grade metamorphism or in regions associated with surficial contact metamorphism. Rocks which formed under regional metamorphic conditions generally show an increase in grain size, with the increase often reflecting the degree of metamorphism.

▪ Foliation

The type of schistosity also provides an indication of the degree of metamorphosis (◘ Table 4.3):

- Low-grade metamorphism—cleavage
- Moderate metamorphism—schistosity
- High-grade metamorphism—gneiss banding

Table 4.2 Source rocks of the various metamorphic rocks

Protolith	Regional metamorphism Low-grade metamorphism (<400 °C)	Medium-grade metamorphism (400–650 °C)	High-grade metamorphism (>650 °C)	Contact metamorphism
Mudstone	Slate/phyllite	Mica schist	Gneiss, granulite, amphibolite	Hornfels
Quartz sandstone	Quartz schist	Quartzite	Quartzite	Quartzite
Graywacke, Arkose	Quartzite, phyllite	Mica schist	Gneiss, granulite	
Limestone	Marble	Marble	Marble	Marble, skarn
Marl, marly limestone	Limestone schist	Mica lime slate	Calcium silicate gneiss	Calc-silicate rock, skarn
Granite, rhyolite	Gneiss	Gneiss	Gneiss	Hornfels
Basalt, diorite, gabbro	Greenschist (low pressure), blueschist (high pressure)	Amphibolite (low pressure), eclogite (high pressure)	Amphibolite, granulite (low pressure), eclogite (high pressure)	Mafic hornfels

Table 4.3 Grain size and schistosity of metamorphic rocks (after Fry 1984)

	Slate group	Gneiss group	Fels group
Grain size	Fine grained	Medium to coarse grained	Fine to coarse grained
Foliation	Strong cleavage, splits into thin sheets	Weaker cleavage, splits into thick plates	No foliation

▪ Mineralogy

Mineralogy contains important information about the nature of the protolith as well as the degree of metamorphism and the metamorphic facies, with specific minerals often related to particular P-T conditions. Examples include: glaucophane (HP or UHP rocks), diamond (UHP rocks) as well as particular so-called index minerals (e.g. chlorite, biotite, hornblende, garnet, staurolite, sillimanite, glaucophane) (Table 4.1).

4.5 Contact Metamorphic Rocks

■ Hornfels

Hornfels (scale 2.2 cm)

Color – speckled appearance, various colors including black, dark brown, gray, and green

Texture – fine to medium grained (sometimes with porphyroblasts)

Structure – often massive, primary structures can sometimes be recognized

Mineralogy – fine-grained matrix with porphyroblasts (e.g., cordierite, andalusite, pyroxene, biotite, garnet, sillimanite)

Occurrence – high-temperature rock, typical in areas of contact metamorphism

Varieties

- **Andalusite-cordierite hornfels**—porphyroblasts of andalusite and/or cordierite
- **Pyroxene hornfels**—porphyroblasts of pyroxene, andalusite and/or cordierite; areas of high-temperature contact metamorphism

▪ Marble

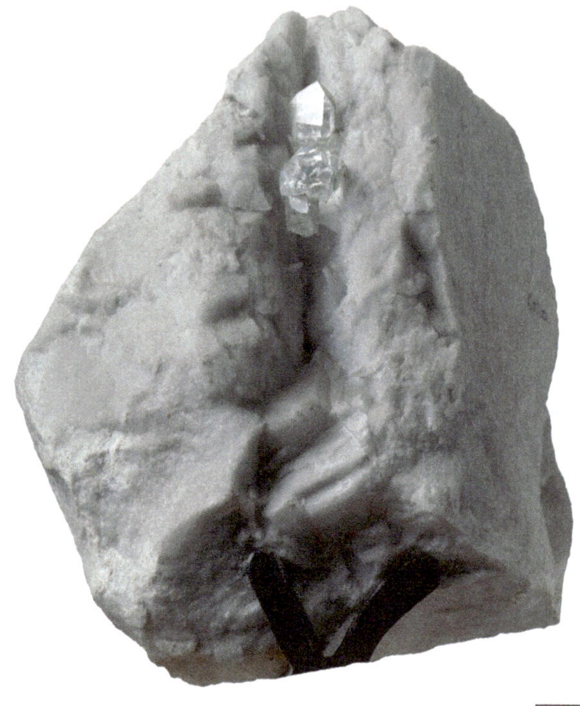

Marble

Color – white or grey, various other colours possible (including black, red, green—often streaked or patchily distributed)

Texture – medium to coarse-grained, granular/sugary

Structure – sedimentary structures (e.g., bedding) may be preserved. Fossils may also be present at low grades of metamorphism

Mineralogy – mainly calcite (up to 99%), also dolomite, sometimes olivine, amphibole, chlorite, serpentine, tremolite, mica, epidote, graphite, plagioclase, pyrite, or quartz

Occurrence – forms due to the metamorphism of limestone around magmatic intrusions. Marble may pass laterally into limestone. Associated with hornfels and skarn

Skarn

Skarn (scale 2.2 cm)

Color – black, brown, gray, often very variable

Texture – fine to coarse grained

Structure – minerals often concentrated in layers, nodules or lenses

Mineralogy – calcium-rich silicates (but also iron, magnesium, manganese, and aluminium silicates), also olivine, garnet, pyroxene, and sulfides

Occurrence – a mineralized calc-silicate rock formed at the contact area between granite (sometimes syenite or diorite) and (mainly) calcareous rocks, as a result of fluid migration (metasomatism); elements (Si, Mg, Fe) from the magma migrate into the calcareous rocks and form a range of silicate minerals and sometimes ores

4.6 Regional Metamorphic Rocks

Slate

Color – various colors, including gray, black, blue, green, and brown

Texture – fine grained

Structure – foliated (slatey cleavage)

Mineralogy – clay minerals (chlorite, kaolinite, illite), quartz, mica (but due to the grain size it is difficult to identify individual minerals);. Sometimes porphyroblasts of pyrite

Occurrence – low-grade metamorphic rock. Regional metamorphic alteration of fine-grained clastic sediments (mudstone, siltstone) or fine-grained tuffs

Phyllite

Phyllite (scale 2.2 cm)

Color – often greenish/grayish, with a characteristic silky sheen on the foliation surfaces

Texture – fine to medium grained. Well developed schistosity (due to the platey minerals); sometimes porphyroblasts

Structure – small crinkles/corrugations often present

Mineralogy – chlorite and/or muscovite (sericite), also quartz, feldspar, biotite, graphite, epidote, and garnet

Occurrence – form due to low-grade metamorphism of pelitic rocks; often pass laterally into mica schist

Mica Schist

Mica schist with garnet (scale 2.2 cm)

Biotite-chlorite schist (scale 2.2 cm)

Color – green or grayish (chlorite schist), gray (graphite schist) or white, (muscovite/sericite schist); brownish or black (biotite schist/muscovite schist). Some mica-rich schists can be shiny/reflective

Texture – fine to medium grained; biotite or muscovite schists can be coarser grained

Structure – lamellar, planar, flakey (schistose)

Mineralogy – usually muscovite and quartz, other minerals (e.g., feldspar, chlorite, muscovite/sericite, biotite, garnet, hornblende, kyanite, sillimanite etc.) may also be present, as may porphyroblasts (e.g., albite in chlorite schist)

Occurrence – forms at higher temperatures than phyllite; fine-grained sediments are the protolith for chlorite schists (low-grade); mica (sericite) schists result when the same rocks undergo a higher grade of metamorphism. Both are often associated with phyllites. Where muddy sandstones are the protolith, quartz-muscovite schists may form

Varieties

- **Biotite schist**—often brown/black, reflective, biotite is an index mineral for regional metamorphism
- **Garnet mica schist**—with garnet porphyroblasts (up to cm size), typical for regional metamorphism
- **Staurolite schist**—high-grade metamorphic rock with staurolite porphyroblasts

4

■ Greenschist

Chlorite schist (scale 2 cm)

Color – green

Texture – fine grained

Structure – schistose, lamellar

Mineralogy – chlorite, epidote, talc, amphibole (actinolite), glaucophane, lesser amounts of quartz and muscovite

Occurrence – protoliths are mainly basalt, gabbro, and also fine-grained sediments

Varieties

- **Talc schist**—soft green slate, very easy to split, feels greasy
- **Chlorite schist**—often green, sometimes albite or chloritoid porphyroblasts
- **Glaucophane schist**—dark colored due to high percentage of glaucophane. A high-pressure metamorphic rock

■ Granulite

Granulite

Color – gray, black, brown, sometimes almost white

Texture – medium to coarse grained, granoblastic

Structure – massive, sometimes weakly banded

Mineralogy – mainly feldspar, sometimes quartz and other minerals (dependent on the protolith, but typically pyroxene, sillimanite, kyanite, garnet, biotite, hornblende). If the source rock is a mudstone, the granulites contain feldspar, quartz, pyroxene, spinel, cordierite, and rare garnets

Occurrence – high temperature/moderate-pressure rocks often formed during regional metamorphism

■ Eclogite

Eclogite (scale 2.7 cm)

Color – green, sometimes reddish

Texture – medium to coarse grained, high density rock

Structure – granular, massive, sometimes banded. Porphyroblasts (garnets, pyroxene) may be present

Mineralogy – pyroxene and garnets. Accessory minerals include rutile, kyanite, hornblende and quartz

Occurrence – form due to the high/ultrahigh pressure alteration of mafic igneous rocks; occurs as blocks/lenses (km size) in masses of igneous/metamorphic rocks; often as xenoliths associated with peridotites and serpentinites

▪ Quartzite

Quartzite (scale 2.2 cm)

Color – white, gray, can also be pink or reddish

Texture – medium to coarse grained

Structure – granular, massive (non-foliated), primary sedimentary structures may be preserved

Mineralogy – quartz, rare feldspar or mica

Occurrence – the protolith is a quartz-rich sandstone; often associated with other metamorphic rocks (e.g., marble, phyllite)

▪ Gneiss

Augen gneiss (scale 17 cm)

Color – alternating light- and dark-colored bands (gneissic banding)

Texture – medium to coarse-grained, light bands are massive, dark bands may show foliation

Structure – compositional banding, with bands > 1 cm; weak foliation

Mineralogy – light-colored bands contain felsic minerals (e.g., feldspar and quartz), while dark-coloured bands contain mafic minerals e.g., mica (muscovite, biotite) and hornblende, sometimes garnet, epidote, pyroxene, sillimanite, and cordierite

Occurrence – high temperature and high pressure rock derived from magmatic (orthogneiss, e.g., granite gneiss) or sedimentary (paragneiss) rocks; associated with both granites/pegmatites and migmatites

Varieties

- **Augen gneiss**—with sheared feldspar porphyroblasts or feldspar/quartz aggregates

Amphibolite

Amphibolite

Color – black, gray, gray-green, dark green, green

Texture – medium to coarse-grained (sometimes fine-grained)

Structure – often massive, foliation may be present (but often weak—due to lack of mica). Sometimes with porphyroblasts

Mineralogy – hornblende, plagioclase, chlorite, epidote, pyroxene, garnet (where $P > 5$ kbar); biotite and rare quartz in para-amphibolites

Occurrence – protolith is often basic igneous rocks (ortho-amphibolite), but can also be sedimentary rocks (para-amphibolite)

▪ Serpentinite

Serpentinite (scale 2.2 cm)

Color – gray-green, green, black. Often irregular bands or flecks of color

Texture – medium to coarse-grained, massive, also fibrous or flaky. Greasy feeling. Sometimes weak schistosity present

Structure – often banded/striped, often crossed by veins of chrysotile serpentine

Mineralogy – serpentine group minerals (chrysotile, antigorite). Accessory minerals include: olivine, pyroxene, amphibole, mica, garnet, chromite, magnetite. Calcite is often present

Occurrence – result from secondary alteration (serpentinization) of ultrabasic magmatic rocks (mainly peridotite). Often in lenses within metamorphic rocks (derived from olivine-rich rocks)

Sedimentary Rocks

Contents

T. McCann, *Pocket Guide Geology in the Field*,
https://doi.org/10.1007/978-3-662-63082-2_5

Sedimentary rocks form at low temperatures and pressures at, or close to, the Earth's surface as a result of the accumulation of particles (**clastic sediments**) or through the precipitation from solutions or the accumulation of organic material (**non-clastic sediments**) (Fig. 5.1). Four groups can be identified:

- **Terrigenous (clastic) sediments**—comprising detrital components (e.g., minerals, lithic fragments, fossil fragments) derived from existing rocks.
- **Biogenic (bioclastic/organic) sediments**—mainly comprising biogenic fragments (e.g., shells) and material formed by organic processes (e.g., peat, coal).
- **Chemical (authigenic) sediments**—formed as a result of direct precipitation of crystalline material from supersaturated fluids (e.g., chert).
- **Volcanoclastic (volcanogenic/pyroclastic) sediments**—comprising mainly detrital materials produced in volcanic areas (e.g. tuff, see ▶ Chap. 3).

5

5.1 Sediment Structures (Current and Wave Movement)

Sedimentary rocks are formed from sediments which have been initially subjected to transport and deposition by water, wind, and ice as well as a result of direct precipitation. The structures that form during deposition are indicative of the energy conditions within the depositing medium (i.e., usually water).

Planar Beds and Laminae

In areas of low energy, for example, when current flow starts to slow down, thin layers of sand are deposited (**parallel lamination**). Laminae have a thickness of <1 cm, while beds are >1 cm thick (Table 5.1).

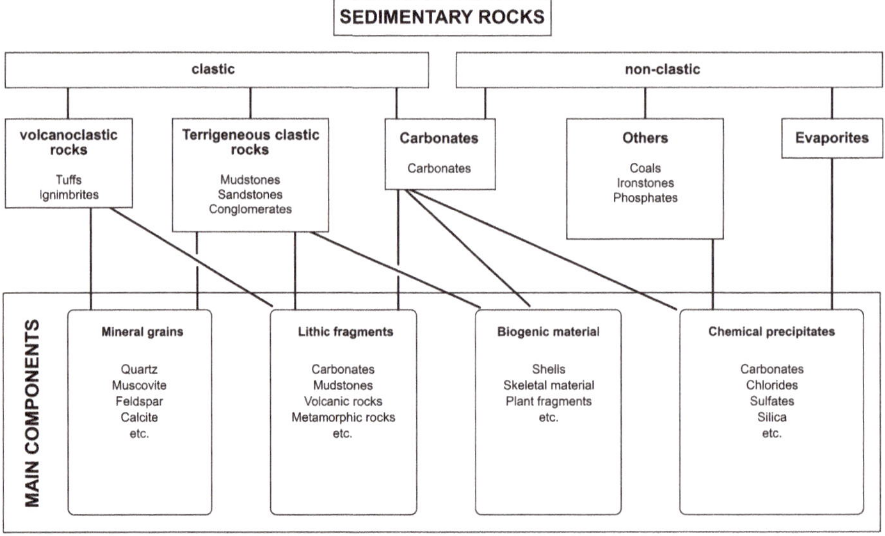

Fig. 5.1 Main components of sedimentary rocks. (After Nichols 2009)

Table 5.1 Form and terminology of layering and lamination. (After Stow 2006)

Designation	Bed thickness (cm)
Bedding	
Very thick bedding	>100
Thick bedding	30–100
Medium bedding	10–30
Fine bedding	3–10
Very fine bedding	1–3
Lamination	
Thick lamination	0.6–1
Medium lamination	0.3–0.6
Fine lamination	0.1–0.3
Very fine lamination	<0.1

Parallel lamination, Almeria, Spain (scale 2.5 cm)

Thick bedding in deltaic sediments, Almeria, Spain (scale 30 cm)

Bedded turbidites, Wales (scale 2.5 cm)

▪ Ripples and Dunes—Cross-Bedding and Cross-Lamination

Ripples are small bedforms that are generated as a result of a unidirectional flow (**current ripple**) or oscillating wave movement (**wave ripple**). While the former are *asymmetrical*, the latter are predominantly *symmetrical*. Dunes are similar to ripples, but are larger bedforms mainly found in marine and aeolian systems and deposited by current or wind action. **Cross-bedding** forms as a result of the downstream migration of ripples or dunes. Planar ripples produce **tabular (planar) cross-bedding**, while wavy ripples produce **trough cross-bedding**.

Current ripple, northern Spain (scale 30 cm)

Cross-bedding/stratification in fossil dunes, Utah, USA (picture width approximately 5 m)

▪ Structures in Sand-Clay Mixtures

Variations in wave or flow energy can result in the formation of intermediate structures with a lenticular or wave-like appearance in mixed-grade sediments. **Flaser lamination** occurs mainly in sand-rich systems, with thin mud laminae/partings. **Lenticular lamination** occurs when isolated ripples are surrounded by mud. **Wavy lamination** is an intermediate form.

▪ Erosional Structures

Erosive currents can result in the formation of a variety of structures in previously deposited sediments (e.g., flute marks/casts or gutter casts).

5

Flute marks, Aberystwyth, Wales (picture width approx. 1 m)

Gutter casts, Harz area, Germany (scale 30 cm)

5.2 Post-Depositional Structures

Recently deposited sediments are often relatively soft due to the amount of water trapped within the pore spaces (Fig. 5.2). This softness/wetness, coupled sometimes with density differences, can result in the formation of a variety of structures:

- Convolute bedding/lamination—deformation occurs due to shock, depositional loading or frictional drag due to the passage of a current.

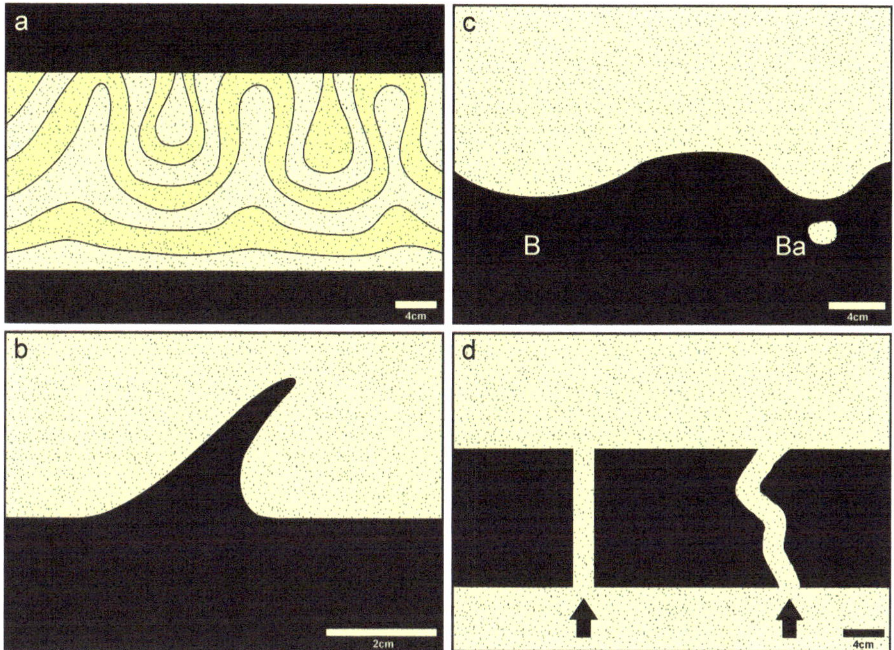

Fig. 5.2 Post-depositional structures in sediments. **a** Convolute bedding. **b** Flame structure. **c** Load structure (B) and ball structure (Ba). **d** Sandstone dykes

- Density contrasts—these can result in denser sediment sinking downwards (load structures, detaching to form ball/pillow structures) or less dense sediment moving upwards (flame structures).
- Sandstone dykes and sills—vertical (dykes) and horizontal (sills) structures which form when liquefied sand is forcefully injected upward through the overlying sediments (often through fractures).
- Dessication cracks (mudcracks)—polygonal cracks which form as a result of the drying out of cohesive muddy sediments in subaerial environments.
- Slides and slumps—mass movements of sediment (up to 500 km^3), which are set in motion by an external trigger (e.g., earthquake, depositional loading) (Fig. 5.2). The material can then move downslope. Slides show no internal deformation, unlike slumps which do.

Dessication cracks, Almeria, Spain (picture width approx. 50 cm)

Slump, Llangranog, Wales (scale 30 cm)

5.3 Biogenic Structures

▪ Ichnofossils (Trace Fossils)

Ichnofossils are records of biological activity which can be described and classified according to their morphology and their position in/on the bed. Ichnofossils and ichnofossil associations (**ichnofacies**) provide much useful information on energy, nutrient, and oxygen levels as well as salinity, and also provide valuable information about water depth.

Ichnofossil: *Scolicia,* Northern Spain (scale 17 cm)

▪ Stromatolites

Stromatolite, Almeria (scale 30 cm)

Laminated (laminae <1 mm, often crinkly/corrugated), structures generally found in fine-grained carbonates. They form as a result of the trapping of carbonate sediment by microbial mats (blue and green algae) and the precipitation of dissolved matter from microorganisms. The resulting structures are often hemispherical in shape.

5.4 Mass Flows

Mass (or gravity/density) flows are mixtures of sediment and liquid that move downslope under the influence of gravity and a variety of physical mechanisms. Four main types can be identified.

▪ Debris Flows

Dense, viscous mixtures of sediment (mud to boulder + size) and water with a higher proportion of sediment than water. The flows are laminar (fragments move parallel to one another; no sediment mixing) and the resulting deposit (deb-
5 rite) is poorly sorted (e.g., matrix-supported conglomerate).

Debrite, Almeria, Spain (scale 30 cm)

▪ Turbidity Flows

Turbulent mixtures of sediment and water (turbidite flows); mostly found in the deep sea. Sediment sorting is very good and forms turbidites with characteristic internal structures (i.e., Bouma sequence, with typically 5 subdivisions).

Turbidite succession, northern Spain (scale 1.8 m)

Grain Flows

Occur as a result of grain-grain collisions in well-sorted sediments (e.g., aeolian sands), where the sediment avalanches down a steep slope.

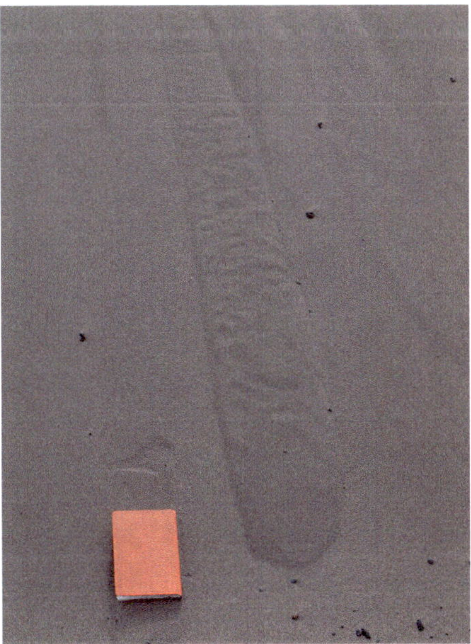

Grain flow, Garzweiler, Germany (scale 20 cm)

- **Liquefied Flows**

Occur when a water-logged sediment body is shocked (e.g., earthquake). This results in the upward movement of pore fluids, resulting in destabilization (liquefaction) of the sediment body.

5.5 Describing Sedimentary Rocks

5.5.1 Clastic Sedimentary Rocks

5

Initial descriptions of clastic sediments can be based on their **composition**—the percentages of the main components—usually quartz, feldspar, and lithic fragments (Fig. 5.3) or the **grain/particle size** of the components.

Based on grain size, clastic rocks can be subdivided into:

- Gravel and conglomerate—grains (clasts) >2 mm in diameter
- Sand and sandstone—grains between 2 mm and 1/16 mm (0.063 mm) in diameter
- Clay, silt, and mudstone, siltstone—grains < 0.063 mm in diameter

The grain size of a sediment depends very much on the energy of the transporting medium, for example, coarse-grained conglomerates require a higher degree of transport energy than fine-grained sands (Fig. 5.4). Other features which are

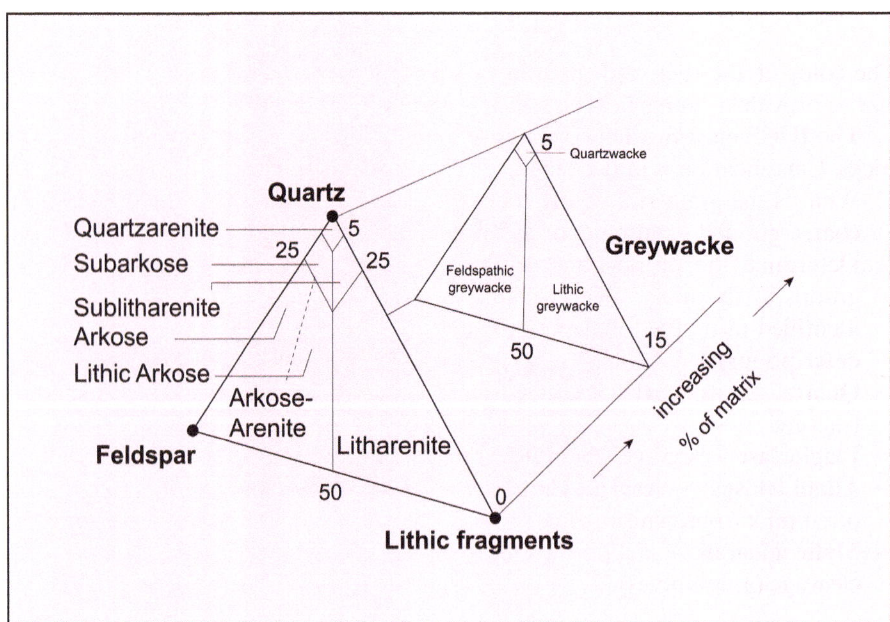

Fig. 5.3 Classification of sandstones. (After Pettijohn 1975)

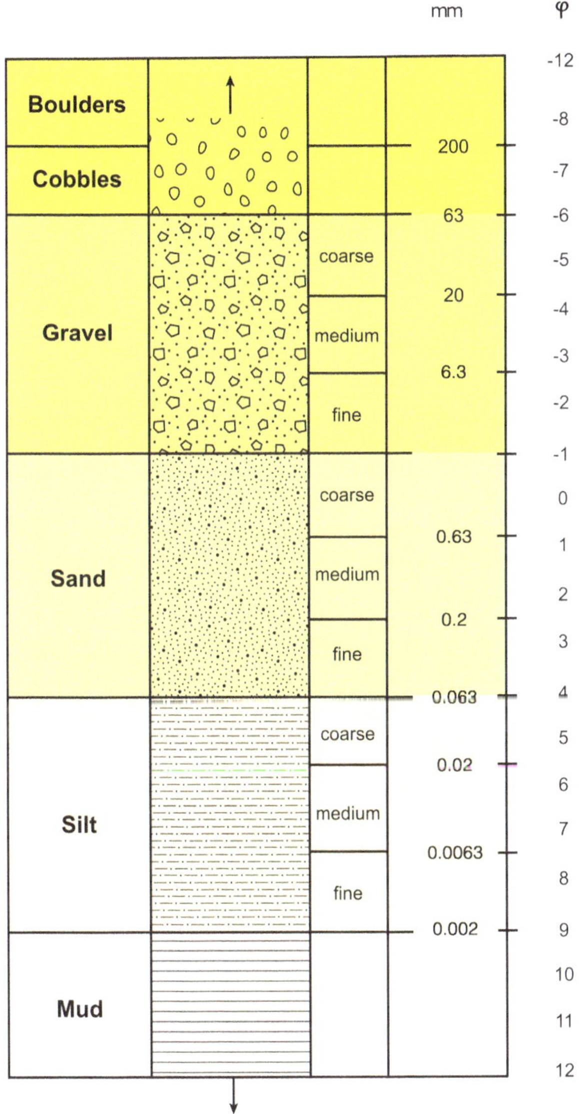

Fig. 5.4 Grain-size distribution and sediment-type classification according to DIN EN ISO 14688 and Udden-Wentworth. (After Stow 2006; Blatt et al. 2006)

used to describe the grains/clasts include, **sphericity** (the degree to which a particle approximates a sphere), **roundness** (the degree of curvature of a particle) and **sorting** (the distribution of grain size, e.g., poorly sorted = range of grain sizes).

Within a bed, grain-size distribution can occur due to increases/decreases in current energy (Fig. 5.5):

5

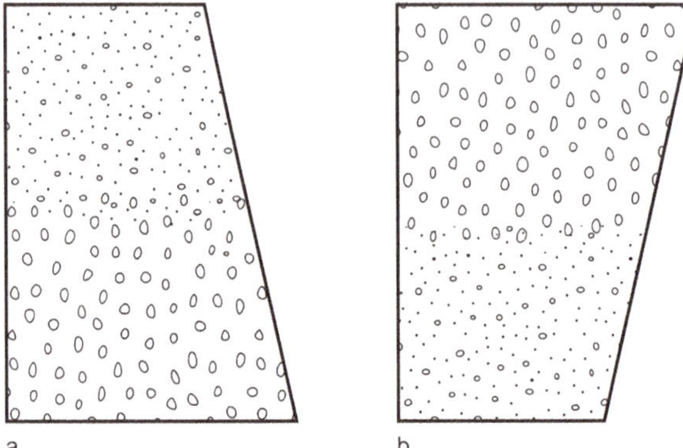

Fig. 5.5 Normal (**a**) and inverse/reverse (**b**) grading in individual deposits. (After Stow 2006; Nichols 2009)

- **normal grading**—upward decrease in grain size (decrease in flow velocity)
- **inverse/reverse grading**—upward increase in grain size (increase in flow velocity)

Normal grading, Uzbekistan (scale 10 cm)

Conglomerates and Breccias

A sedimentary rock comprising a mixture of clasts (diameter >2 mm) in a finer-grained matrix. A conglomerate contains rounded clasts, whereas a breccia has angular clasts (Fig. 5.6).

- **Matrix-supported conglomerates**—the individual clasts are separate from one another
- **Clast-supported conglomerates**—the individual clasts are in contact with one another

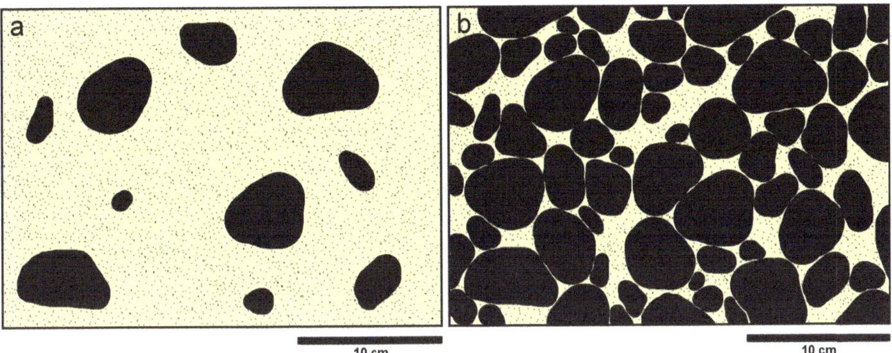

Fig. 5.6 Description of the distribution of clasts within a conglomerate. **a** Matrix-supported conglomerate. **b** Clast-supported conglomerate

Two main subdivisions are made based on composition:

- **monomictic**—the clasts all have the same lithology
- **polymictic**—the clasts comprise a range of lithologies

Sandstones

Sands and sandstones are sediments where the individual grain sizes range from 0.063–2 mm. They can be subdivided according to their grain size into very fine, fine, medium, coarse and very coarse (see Fig. 5.4). The percentage of matrix or feldspars present can also be used to classify sandstones (e.g., greywacke, arkose).

Silts and Muds

The most fine-grained sediments, mainly deposited in low-energy environments, are silts (0.002–0.063 mm) and clays (<0.002 mm). The former can be divided into coarse, medium, fine, and very fine (Fig. 5.4). Muds are mixtures of silt- and clay-sized particles.

5.5.2 Non-Clastic Sedimentary Rocks

Limestone and dolostone (dolomite) Carbonate rocks and sediments are predominantly of biogenic origin (e.g., shells and shell fragments, calcareous algae) forming as a result of carbonate precipitation, coupled with biochemical processes (Table 5.2). By definition, a **limestone** is any sedimentary rock containing >50% calcium carbonate, whereas dolostone (dolomite) is a carbonate rock with a high magnesium content. Carbonate production is mainly (but not exclusively) limited to tropical and subtropical areas. Most limestones are formed in coastal areas and shallow-marine environments. In addition, carbonates also form in caves (stalagmites/stalactites), springs (tufa), soil horizons (calcrete), lakes, and in deep-marine environments.

Table 5.2 Classification of carbonate rocks. (After Stow 2006)

Main types	Subtypes	Properties and genesis
Limestones ($CaCO_3$ dominant)	Subdivision based on: a) Grain size: calcirudite >2 mm calcarenite 0.063–2 mm calcilutite <0.063 mm b) Textural features (according to Dunham): grainstone, packstone wackestone, mudstone, boundstone, floatstone, rudstone, bafflestone, bindstone, framestone	Carbonate particles formed by primary chemical precipitation, biogenic secretion, and as fragments of limestone skeletons. Also erosion of existing carbonate rocks
Dolomite ($Ca\,Mg(CO_3)_2$)	Subdivision according to the degree of dolomitization: <10% dolomite = limestone; 10–50% dolomite = dolomitic limestone; 50–90% dolomite = calcareous dolomite; >90% dolomite = dolomite	Most dolomites/dolostones are formed by partial or complete displacement of limestone; may occur during early diagenesis in evaporitic environments, but usually form subsequent to burial

5

Calcium carbonate can occur either as **calcite** (stable form) or **aragonite** (unstable at Earth surface temperatures and pressures). Aragonite recrystallizes over time to calcite. The calcium ion in $CaCO_3$ is sometimes replaced by other minerals, e.g., Mg (dolomite), Sr (strontianite) and Fe (ankerite).

Texturally, carbonates are similar to clastic sediments (e.g., grain roundness, grain sorting) or to chemical precipitates (crystal intergrowths), or they may be a mixture of both. In addition, they also include a variety of organic structures (e.g., stromatolites, reefs). Based on their grain size, carbonates can be divided into three main types:

- calcirudite (equivalent to conglomerate)—grains (clasts) >2 mm in diameter
- calcarenite (equivalent to sandstone)—grains between 0.063–2 mm in diameter
- calcilutite (equivalent to mudstone)—grains <0.063 mm in diameter

More detailed schemes are also available, one is described below.

Dunham Classification

This classification scheme uses mainly textural criteria, although the nature of the grains (e.g., ooid, peloid) can also be taken into account. It is mainly used to describe carbonates in outcrop/hand sample (Fig. 5.7).

The main components are skeletal fragments (e.g., molluscs, brachiopods, sponges, corals) and carbonate-forming algae (e.g., red and green algae) as well as bioherms (e.g., corals, stromatolites). Non-biogenic components are also important (Fig. 5.8), including

Components	Support / grain content	Criterion	Name
Components not bound together during deposition	mud supported	less than 10% grains	Mudstone
		more than 10% grains	Wackestone
	grain supported		Packstone
	lacks mud; grain supported		Grainstone
components bound together during deposition			Boundstone
Depositional texture not recognizable			Crystalline
Components not bound together during deposition	> 10% grains > 2mm	matrix supported	Floatstone
		supported by components > 2 mm	Rudstone
Components bound together during deposition	Organisms trap sediment		Bafflestone
	Organisms bind sediment		Bindstone
	Organisms provide a framework		Framestone

Fig. 5.7 Classification of limestones based on depositional texture. (After Dunham 1962, with modifications by Embry and Klovan 1971)

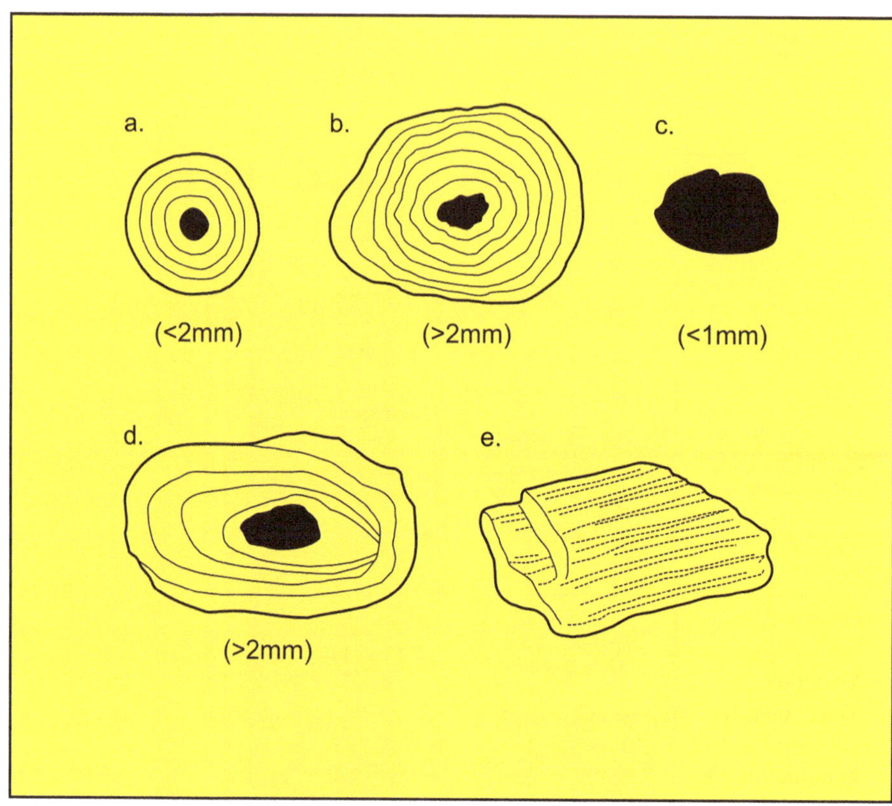

Fig. 5.8 Non-biogenic fragments in limestones: **a** Ooid. **b** Pisoid. **c** Peloid. **d** Oncoid. **e** Intraclast. (After Nichols 2009)

- Ooids—small (<2 mm in diameter), spherical carbonate grains with a concentric internal structure. Form in moderate to high-energy environments.
- Peloids—small (<2 mm in diameter), spheroidal, ovoid or irregularly-shaped carbonate grains with no internal structure. Often the faeces of marine organisms.
- Oncoids—roughly spherical grains/clasts (up to 10 cm diameter) often formed by encrusting cyanobacteria.
- Intraclasts—carbonate fragments that have been eroded, transported, and re-deposited within the depositional environment.

Table 5.3 Main minerals in evaporite deposits. (After Stow 2006; Blatt et al. 2006)

Main types	Minerals	Chemical composition
Marine evaporites		
Chlorides	Halite	NaCl
	Sylvite/Sylvine	KCl
Sulfates	Anhydrite	$CaSO_4$
	Gypsum	$CaSO_4\ 2H_2O$
Carbonates	Calcite	$CaCO_3$
	Dolomite	$CaMg(CO3)_2$
Nonmarine evaporites		
Carbonates	Trona	$NaHCO_3\ Na_2CO_3\ 2H_2O$
Sulfates	Gypsum	$CaSO_4\ 2H_2O$
	Anhydrite	$CaSO_4$
Chlorides	Halite	NaCl

5.5.3 Evaporites, Cherts, Iron-Rich Rocks, Phosphates and Organic Sediments

Evaporites are formed by precipitation from an aqueous solution which, as a result of evaporation, has become concentrated with respect to a particular mineral. Around 70 different evaporite minerals are known (Table 5.3).

Evaporites are often deeply buried (e.g., Zechstein, Northern Germany), and many of these ancient evaporite deposits are thick (100's–1000's m).

Cherts are fine-grained siliceous sedimentary rocks consisting of microcrystalline (or cryptocrystalline) quartz crystals (Table 5.4). Cherts are mainly of biological origin (i.e., the skeletons of microorganisms) or they form due to the replacement of existing material (e.g., petrified wood).

Ironstones are deposits with at least 15% iron content. Two main types are known:

- Precambrian banded-iron formations
- Iron ooliths

Carbon-rich **organic rocks** are of great economic value. Two main forms occur—**coal** and **oil shale**.

Table 5.4 Classification of of siliceous sediments. (After Stow 2006)

Main types	Subtypes	Properties and genesis
Bedded chert	Radiolarian chert Diatomaceous chert Sponge spicule chert Jasper	Mostly of marine origin; comprising recrystallized quartz and biogenic residues
Nodular chert	Flint	Bulbous chert, often in writing chalk
	Silcrete	Nodular/encrusting chert formed in certain soils or as surface crusts
Partly-lithified siliceous sediments	Radiolarite	Rich in radiolarians
	Diatomite	Rich in diatoms
	Spiculite	Rich in sponge spicules

5.6 Selected Sedimentary Rocks

▪ Conglomerate

Conglomerate (scale 30 cm)

Color – variable—dependent on components present

Texture – rounded clasts (>2 mm diameter) in a finer-grained matrix

Structure – structureless or graded/bedded. Both clast- and matrix-supported conglomerates can occur

Mineralogy – polymictic (several clast types), monomictic (one clast type)

Occurrence – indicative of high-energy flows (e.g., clast-supported conglomerates in fluvial environments) or debrites (e.g., matrix-supported conglomerates)

Breccia

(scale 2.5 cm)

Color – variable—dependent on components present

Texture – angular clasts (>2 mm diameter) in a finer-grained matrix

Structure – structureless or graded/bedded. Both clast- and matrix-supported breccias possible

Mineralogy – polymictic (several clast types), monomictic (one clast type).

Occurrence – indicative of flows with high energy. Often close to the source area (lack of clast rounding implies relatively short transport distance)

▪ Sandstone

Sandstone (scale 2.2 cm)

Color – variable—often red, brown, greenish, yellow, gray, or white

Texture – fine to coarse grained (0.063 to max. 2.0 mm diameter). Fragments are angular to well rounded and in a fine-grained matrix/cement

Structure – structureless or graded/bedded (possibly with a range of internal structures)

Mineralogy – quartz, feldspar, and rock fragments

Occurrence – indicative of medium- to higher-energy depositional environments

▪ Arkose

Arkose (scale 1 cm)

Color – red, pink

Texture – fine to coarse-grained (max. 2.0 mm diameter). Mainly quartz and feldspar fragments (25–50%) in matrix/cement. Fragments are angular to well rounded

Structure – structureless or graded/bedded (possibly with a range of internal structures).

Mineralogy – quartz and feldspar—often a weathering product of granitoids

Occurrence – indicative of medium- to higher-energy depositional environments

▪ Greywacke

Greywacke (scale 2 cm)

Color – grey, black

Texture – fine to coarse-grained (max. 2.0 mm diameter) fragments in a fine-grained matrix (up to 15%). Fragments are angular to well rounded

Structure – structureless or graded/bedded (possibly with a range of internal structures, e.g., Bouma sequence)

Mineralogy – quartz, feldspar and rock fragments

Occurrence – indicative of medium- to higher-energy flows (e.g., turbidity flows)

Siltstone

5

Siltstone beds in mudstone (scale 16 cm)

Color – variable—black, gray, brown, yellow, white

Texture – fine-grained (0.002–0.063 mm diameter). Quartz and feldspar fragments in matrix/cement. Grains are angular to well rounded

Structure – structureless or graded/laminated/bedded

Mineralogy – quartz and feldspar

Occurrence – indicative of low-energy environments

Mudstone

Mudstone (scale 2 cm)

Color – variable—black, gray, brown, yellow, green, red, white

Texture – very fine grained. Individual grains are not visible

Structure – structureless or laminated

Mineralogy – mainly clay minerals, as well as quartz and feldspar

Occurrence – indicative of flows with very low energy. Mudstones are mixtures of both fine-grained silts and clays. The degree of silt-richness can be determined by chewing gently on a small sample (crunchy = silty)

Limestone

Bioclastic limestone (scale 2.2 cm)

Color – pure limestone is often gray, white, or creamy, impure limestone can be red, brown, or black

Texture – variable and dependent on fossil content and grain size. Range from fine-grained (micrite/chalk) to crystalline (sugar-like appearance) to conglomeratic or brecciated limestones

Structure – structureless or graded/bedded (possibly with a range of internal structures). Large structures (e.g., reef bodies) are sometimes well developed

Mineralogy – mainly calcite. Other minerals may be present (e.g., quartz—as grains or in the form of chert)

Occurrence – biochemical rocks comprise the remains of organisms (shells, skeletons), carbonate clasts (intraclasts), or other fragments (ooids, peloids). Usually shallow marine and above the carbonate compensation depth (c. 4200–5000 m). Often in combination with other limestone types

Varieties

- **Shelly limestone**—limestones with high proportions of shells/shell fragments

- **Oolitic limestone**—ooids are spherical grains of calcium carbonate (<2 mm in diameter), which are formed by precipitation in high-energy shallow marine areas
- **Chalk/micrite**—a fine-grained limestone. Chalk is formed from microfossils (coccoliths) in deep-marine areas. Micrite is a general term for fine-grained limestone
- **Travertine/Tufa/Dripstone**—a fine-grained freshwater limestone, often found in caves (**stalactites**—growing from the ceiling, **stalagmites**—growing from the floor)

Dolostone (Dolomite)

Dolostone (dolomite) (scale 2.2 cm)

Color – white, creamy, gray

Texture – dolostone is sometimes primary, but usually secondary (i.e., a diagenetically altered limestone). Textures are sometimes like those of the original rock, but can also be very altered (granular to fine grained)

Structure – structureless to bedded. Internal structures are sometimes present

Mineralogy – mainly dolomite. Sometimes with calcite or quartz/chert

Occurrence – often associated with limestones

Chert (Flint)

Flint (scale 1 cm)

Color – gray, black

Texture – fine-grained, with conchoidal fracture

Structure – tuberous, sometimes layered

Mineralogy – microcrystalline quartz

Occurrence – often associated with limestone. Particular types include diatomite (silica from diatom skeletons) or radiolarite (silica from radiolaria skeletons)

Peat

Peat (scale 1 cm)

Color – brown, black

Texture – plant remains can be clearly seen

Structure – bedded/layered

Mineralogy – organic material (approx. 60% C), plant remains

Occurrence – marshes/swamps, associated with other sediments

▪ Anthracite

Anthracite (scale 2 cm)

Color – black, often metallic shiny

Texture – fine grained

Structure – layered (seams)

Mineralogy – carbon (approx. 94%), rare plant remains

Occurrence – very pure form of coal. Often associated with other sedimentary rocks. Found in continental depositional environment

Fossils and Paleoecology

Contents

T. McCann, *Pocket Guide Geology in the Field*,
https://doi.org/10.1007/978-3-662-63082-2_6

Fossils are the preserved, fossilized remains or traces of animals, plants, and other organisms. They are mainly found in sedimentary rocks, particularly in limestones, mudstones, siltstones, and sandstones. Biostratigraphy is an important tool for the correlation and relative dating of rock units or successions. Certain stratigraphic periods are typified by characteristic fossil groups (i.e., **index fossils**).

The main controlling mechanisms for the distribution of recent (and fossil) organisms include:

- Temperature (affects large-scale distribution, both numbers and diversity)
- Light (the range of maximum biological productivity is in the upper 10–20 m of the water column)
- Oxygen levels
- Substrate (e.g., mud, sand)
- Salinity
- Water turbulence
- Nutrients
- Climate (e.g., tropical, temperate)

The marine environment, which is of particular importance for organisms, can be subdivided based on water depth (Figs. 6.1 and 6.2). Organisms living on the sea floor are termed **benthic**, while those living within the water column can be divided into two groups:

- **Planktic** (= **planktonic**), organisms that are non-motile (i.e. drifting in the water column) or are weak swimmers, and
- **Nektic** (= **nektonic**), organisms which are strong swimmers.

Neritic organisms live in shallow coastal waters, while **pelagic/oceanic** faunas inhabit the surface waters or mid-depths of the open oceans.

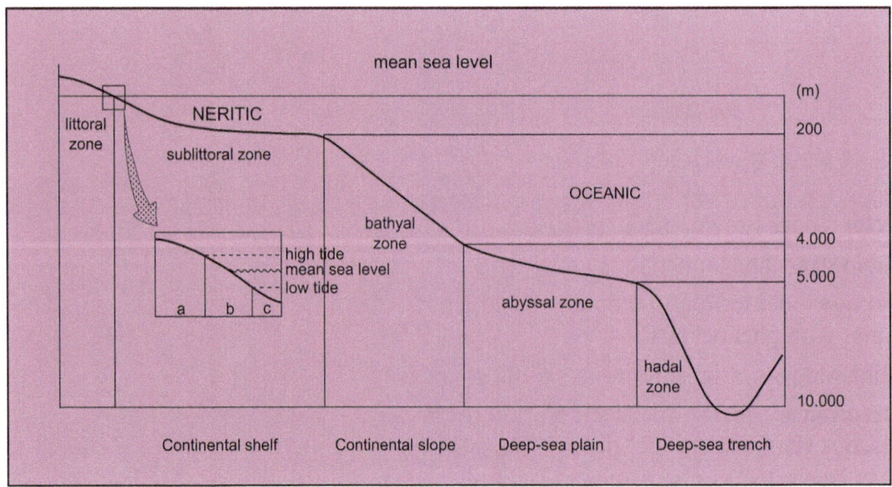

Fig. 6.1 Marine environments, **a** supralittoral, **b** littoral, **c** sublittoral. (After Clarkson 1998)

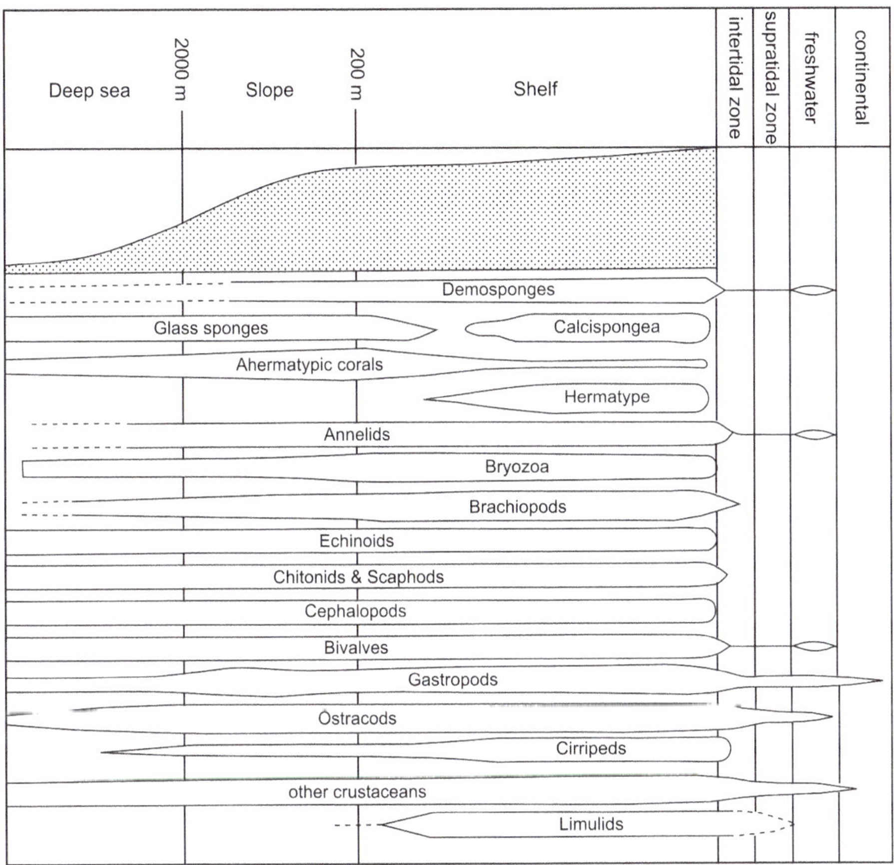

Fig. 6.2 Distribution of living organisms according to water depth. (After Benton and Harper 2009)

6.1 Fossil Groups

▪ Sponges (Porifera)

Sponge *Coeloptychium deciminum* (Cretaceous)

Period – Cambrian—Recent

Milieu – subaquatic (marine—fresh water, polar—tropical, shallow marine—abyssal)

Sponges are nodular to goblet-shaped organisms, sedentary filter-feeders that pump water through internal channel systems.

▪ Cnidarians (Cnidaria, class Anthozoa: corals)

Rugose Coral: *Hexagonaria hexagona* (Middle Devonian)

Scleractinian coral: *Favia* (Recent)

Tabulate coral: *Favosites basaltica* (Devonian)

Period – Ordovician—Recent

Milieu – marine, shallow (benthic) to greater depths (e.g., rugose corals)

This group comprises a wide range of both solitary and colonial cnidarians, including jellyfish, sea anemones, and corals. The most important cnidaria fossils are corals because of their reef-building function, although soft tissue Cnidaria (without skeletons) are known from the Neoproterozoic (Ediacaran fossils). In general, coral diversity decreases with increasing water depth. Corals have massive, external, calcareous skeletons. Four main groups have been identified:

- Tabulate corals (Order Tabulata)—exclusively colonial corals comprising individual cells/corallites (Lower Ordovician—Permian)
- Rugose corals (Order Rugosa)—solitary and colonial corals (Middle Ordovician—Upper Permian)
- Scleractinian corals (Order Scleractinia)—solitary and colonial corals (Middle Triassic—Recent)
- Octocorallia (Subclass Octocorallia)—colonial organisms, both soft bodied and with internal skeletons (? Ediacara—Recent)

- **Molluscs (Mollusca)**

Period – Cambrian—Recent

Milieu – marine, freshwater and terrestrial

Gastropod, *Murchisonia binodosa* (Middle Devonian)

Ammonite, *Perisphinctes* sp. (Jurassic)

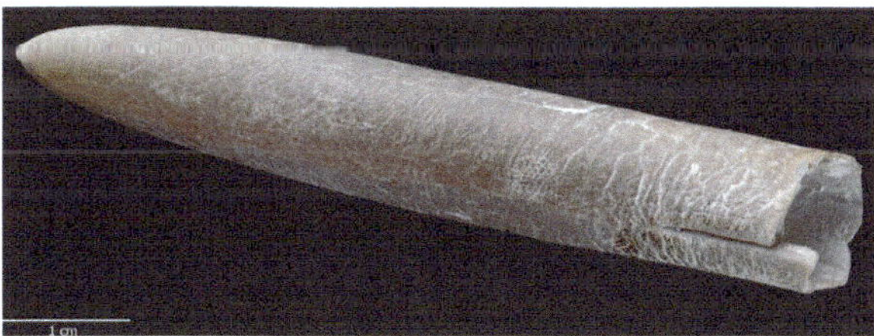

Belemnite, *Belemnitella mucronata* (Cretaceous)

Nautilus (Recent)

Pelecypod/Bivalve *Mercenaria mercenaria* (Pliocene)

This group contains a broad range of organisms, including: snails, mussels, oysters, squids, and octopuses among others. Their initial appearance in the Cambrian was followed by a marked radiation in the Ordovician. Three main groups are distinguished:

- Gastropods (Class Gastropoda)—molluscs, in which the soft body is mostly covered by an asymmetrical, spirally-coiled shell (Cambrian—Recent).
- Cephalopods (Class Cephalopoda)—bilaterally symmetrical, with straight, curved or coiled shells, which are usually external, and that show a variety of forms. Cephalopods are the largest, most intelligent, and agile of the molluscs (Cambrian—Recent). Within the cephalopods, three groups are of great stratigraphic importance:
  - Nautiloids (Subclass Nautiloidea)—diverse group of small to large cephalopods (Cambrian—Recemt),
  - Ammonites (Subclass Ammonoidea)—diverse group of small to large (adult animals could reach up to 6 m in diameter) cephalopods with external shells, which are usually planispiral (Devonian—Cretaceous),
  - Belemnites (Order Belemnitida)—bullet-shaped cephalopods with an internal skeleton (Carboniferous—Cretaceous).
  - Pelecypods/Bivalves (Class Pelecypoda)—a morphologically highly diverse group with shells where the individual valves are mirror images of each other (bilaterally symmetrical). Shell forms range from typical bivalves (where both shells are the same size, and the line of symmetry runs along a plane between both valves; the individual valves are generally asymmetrical) to large horns (e.g., rudists, which were the most important reef builders in the Upper Cretaceous) and elongated cylindrical tubes (e.g., razor clams) (Cambrian—Recent).

Brachiopods (Brachiopoda)

Brachiopod (Articulata), *Coenothyris vulgaris* (Muschelkalk)

Brachiopod (Inarticulata), *Lingula* sp. (Upper Devonian)

Period – Cambrian—Recent

Milieu – mainly shallow marine (some with high salt tolerance), some species are found at depths of up to 6000 m

A diverse group of shelled organisms which are of great stratigraphic importance, especially in the Paleozoic. The two valves, which are usually of different sizes, generally have a plane of symmetry through the middle of the valve, and so can be distinguished from the Pelycypods. There are two main groups:

- Class Inarticulata—brachiopod group which were especially important in the Lower Paleozoic (most Families were extinct by the end Devonian),
- Class Articulata—the most important brachiopod group.

▪ Echinoderms (Echinodermata)

6

Sea lily, *Encrinus liliformis* (Middle Triassic)

Sea urchin, *Cidaris vendocinensis* (Cretaceous)

Period – ?Neoproterozoic—Recent
Milieu – marine (shallow—abyssal)

A diverse group in which the individuals generally have a five-sided (pentaradial) symmetry. Various classes are identified, including:

- Sea urchins (Class Echinoidea)—rounded-, pentagonal- or heart-shaped organisms comprise this subgroup, which is of great stratigraphic importance (Lower Cambrian—Recent),
- Sea lilies (Class Crinoidea)—Echinoderms with a conical, spherical, or goblet shaped central body, bearing five or more arms. The body may be attached to a long, columnar stem, or the crinoids may be free swimming (middle Cambrian—Recent),
- Blastoids (Subphylum Blastozoa, Class Blastoidea)—stemmed echinoids, often stalked and with a conical or spherical body (Silurian—Permian),
- Cystoids (Subphylum Blastozoa, Class Eocrinoidea)—the oldest arm-bearing echinoderms, similar in form to crinoids but the main body is more oval in shape (Cambrian—Silurian),

Graptolites (Hemichordata)

Graptolites: *Didymograpus sp.* (Ordovician)

Period – Cambrian—Upper Carboniferous

Milieu – marine

Graptolites are the fossil remains of colonial, planktonic to benthic organisms. They are of great stratigraphic importance in the Lower Paleozoic.

Arthropoda

The arthropoda are a large and diverse group that includes insects, crabs, shrimps, and ostracods.

Trilobites (Class Trilobita)

6

Trilobite *Dalmanites* sp. (Lower Silurian)

Period – Cambrian—Permian

Milieu – shallow marine

Trilobites have a dorsal carapace and their bodies can be subdivided into three parts: cephalon (head), thorax (trunk), and pygidium (tail). Trilobites are characteristic of the Cambrian (maximum diversity is in the Upper Cambrian), with their occurrence decreasing noticeably in the Ordovician. They had complex compound eyes, but could also be blind. They are a morphologically diverse group, reflecting the degree of adaptation to their environment (e.g. changes in salinity, temperature, sediment type).

Supplementary Information

T. McCann, *Pocket Guide Geology in the Field*,
https://doi.org/10.1007/978-3-662-63082-2

References

Benton MJ, Harper DAT (2009) Introduction to paleobiology and the fossil record. Wiley-Blackwell, Oxford, S 592

Blatt H, Tracy RJ, Owens BE (2006) Petrology. Igneous, sedimentary, and metamorphic. New York, W. H. Freemann and Company, S 530

Clarkson ENK (1998) Invertebrate palaeontology and evolution. Blackwell Science, Oxford, S 452

Coe AL (2010) Geological field techniques. Wiley-Blackwell, Oxford, S 323

Dunham RJ (1962) Classification of carbonate rocks according to depositional texture. In: Ham WE (Hrsg) Classification of carbonate rocks, 1st ed. Memoir, American Association of Petroleum Geologists. American Association of Petroleum Geologists, Tulsa, S 108–121

Embry AF, Klovan JE (1971) A late Devonian reef tract on north-eastern Banks Island, Northwest Territories. Bull Can Pet Geol 19:730–781

Hamilton WR, Woolley AR, Bishop AC (1974) The Hamlyn guide to minerals, rocks, and fossils. Hamlyn, London, S 320

Jerram D, Petford N (2011) The field description of igneous rocks, 2nd ed. Geological field guide series. Wiley-Blackwell, Oxford, S 238

Markl G (2004) Minerale und Gesteine. Eigenschaften-Bildung-Untersuchung. Elsevier & Spektrum Akademischer, Heidelberg, S 355

McCann T, Valdivia Manchego M (2015) Geologie im Gelände. Das Outdoor-Handbuch. Springer Spektrum, Heidelberg, S 376

Nichols G (2009) Sedimentology and stratigraphy, 2nd ed. Wiley, Oxford, S 432

Orton GJ (1996) Volcanic environments. In: Reading HG (Hrsg) Sedimentary environments: processes, facies and stratigraphy. Blackwell Science, Oxford, S 485–567

Pettijohn FJ (1975) Sedimentary rocks, 3rd ed. Harper & Row, New York, S 628

Philpott AR, Ague JJ (2009) Principles of igneous and metamorphic petrology. Cambridge University Press, Cambridge, S 684

Ronov AB, Yaroshewsky AA (1969) Chemical composition of the Earth's crust. In: Hart PJ (Hrsg) The Earth's crust and upper mantle. American geophysical union, Washington D.C., S 37–62

Schumann W (2007) Der große Steine- und Mineralienführer. BLV Buchverlag, München, S 399

Stow DAV (2006) Sedimentary rocks in the field. A colour guide. Academic Press, San Diego, S 320

Thorpe R, Brown G (1985) The field description of igneous rocks. Geological society of London handbook. Wiley, Chichester, S 155

Wenk H-R, Bulakh A (2004) Minerals. Their constitution and origin. Cambridge University Press, Cambridge

Related Link

OutcropWizard (▶ www.outcropwizard.de)

GPSR Compliance

The European Union's (EU) General Product Safety Regulation (GPSR) is a set of rules that requires consumer products to be safe and our obligations to ensure this.

If you have any concerns about our products, you can contact us on ProductSafety@springernature.com

In case Publisher is established outside the EU, the EU authorized representative is:

Springer Nature Customer Service Center GmbH
Europaplatz 3
69115 Heidelberg, Germany

Batch number: 10370712

Printed by Printforce, the Netherlands